Religion, Neuroscience and New Physics in Dialogue

Religion, Neuroscience and New Physics in Dialogue

Stone Age Souls in Modern Minds

Darren C. Marks

ANTHEM PRESS

Anthem Press
An imprint of Wimbledon Publishing Company
www.anthempress.com

This edition first published in UK and USA 2021
by ANTHEM PRESS
75–76 Blackfriars Road, London SE1 8HA, UK
or PO Box 9779, London SW19 7ZG, UK
and
244 Madison Ave #116, New York, NY 10016, USA

British Library Cataloguing-in-Publication Data
A catalogue record for this book is available from the British Library.

Library of Congress Control Number: 2021942254

ISBN-13: 978-1-83998-009-1 (Hbk)
ISBN-10: 1-83998-009-5 (Hbk)

Cover Image: Painting "Fallen Man (Back Panel)" by Kai Wolfram Von Maltzahn (date unknown)
Oil on plywood, original 24 x 34 inches, private collection

This title is also available as an e-book.

"Civilization has little to fear from educated people and brain workers"

S. Freud

"Realism is the only philosophy that doesn't make the success of science a miracle"

H. Putnam

"The saddest aspect of life right now is that science gathers knowledge faster than society gathers wisdom"

I. Asimov

CONTENTS

PREFACE

> As long as the World is turning and spinning, we are gonna be dizzy and make mistakes.
>
> Mel Brooks

Strange days. I sit in my home office, using Zoom to lecture and to communicate, while a global pandemic brings about a second lockdown in most European countries (and some of my province in Canada) while I listen to the national radio broadcast a "science" reporter whose expertise a short 10 months ago was provincial governmental affairs. In the period of the pandemic, he went from being an expert on politics to the public radio voice of the science of the pandemic on the national radio broadcast in my heavily populated region. His position on the national broadcaster, its centrality in terms of Ontario population, means that he really is shaping how many educated or interested Ontario residents think about the pandemic. He is, in the parlance I will repeatedly use in this work, "educating us to reality"—in this case not only on the science of COVID-19 but also the human ethics and social aspects of it as well. Our intrepid reporter cites health experts and various studies, interviews talking heads from various universities with impressive expertise and works tirelessly but gets so much of it wrong (at least, for me)—I checked his biography and as far as I can see, he has no training in science. The talking heads struggle to speak "science" into a void that wants easy reduction and whenever they qualify or modulate what can be said, end up being pressed to an absolute kind of statement on numbers, virulence or infection rates as if they have absolute secret knowledge that is fated to come. And importantly, each morning the political reporter qua science expert tells the audience how that understanding should impact or alter human behavior, social interactions and even economics or politics based on his curious interactions with above experts. But, and this is the point, he confuses the two places on what science is saying and what it means. That is an intractable point of confusion, a place of dialogue between the scientists and the general public for whom public agencies are working. Our education to reality in this pandemic in my country is a confused mess of science, information, social and economic impacts dictated by emotion as much as by reason or logic. It strikes me that all of this is the purpose of this book in a nutshell. A reflection on a world that "speaks" scientifically but applies that narrative from perspectives that science would barely recognize as being scientific. The world of COVID-19 is a world of opinion shared on social media with all kinds of PhDs gleaned from Google University, all speaking at times the real science of virology, epidemiology and medicine while interpreting this into the moral and social world of public health, social behavior and individual rights. We are at once "speaking" science but really talking about ourselves. And, as I note in the first chapter, in the greater context of everything humans may do in the future regarding climate change, this is a very old issue of the confusion of what science and human behavior based on that claim of reality may be. It may end up costing us everything if we get it incorrect.

Originally, this work was to be the work of two minds. I prefer collegial interdisciplinary work as it allows me to learn much as our intrepid reporter friend has over the last year of the pandemic. I found that at Western University there was one of the most brilliant neuroscientists in the field working at a very large lab who shared a similar background as me in the Oxbridge experience and like me, remembered fondly the shared senior common rooms of faculty from different fields rather than the siloed faculties of the North American university. He and I lunched, chatted about his work and mine. We talked about neuroscience, physics, philosophy, and our families. Quickly, however, I found that when I talked about the ideas of neuroscience that interested people in the religious or philosophical worlds—consciousness, identity, selfhood, agency and so forth—this extraordinarily brilliant man looked at me as if I was speaking Martian. In his laboratory and work, these ideas were not a part of the discussion of what he or his group was doing. They might think about them from time to time, but they belonged to another field, and he would laugh and tell me "your field." There was and is modesty in his work, a focus on what is reproducible, scientific in method and applicable in terms of the physics, biology or chemistry that was at work. His research group, inaptly named "mind and brain" left the "mind" to psychologists and philosophers while he was interested in the brain. This was the small clump of biomatter that he could measure events within, either electrical or chemical. He gave evidence to governmental groups on what it meant to be brain dead, what it meant to be "locked down" but he did not make the judgment of what that might mean legally, morally or socially. I realized that the "mind" was not of study for him, it was all brain. It is not for me to report what he thought might be the metaphysics of simply brain, the reduction of mind to biomass all the way down as he did think of these things, but the point is that he realized there was a limit of what he, qua scientist, could say about these things before he left physics and entered metaphysics.

Unlike my radio expert, he realized that the end of science and the beginning of human application in all its messy manifestations was a cul de sac. It did have meaning, but he would, as I will point out in the chapter on physics, simply "shut up and calculate"—namely said modest things about the nature of the Real in the brain. But isn't that the confusion? That we cannot simply remain as "shut up and calculate" but must make meaning? Science is not just how a virus spreads, how it infects, how it overwhelms an immune system but also how humans organize themselves around this knowledge. How humans make judgments, even moral ones, on who is mask wearing, responsible or irresponsible? How we reduce this scientific truth of a virus into real and religious politics—from which are essential businesses to conspiracy theories of QAnon, anti-vaxxer attitudes—and how some religious communities, especially in Canada, the United Kingdom and the United States almost default to an anti-science, pro-democracy or freedom attitude. Anti-maskers, pro-conspiracy, climate deniers and anti-science people shout about Jesus as their personal savior while decrying Satanic, antisemitic conspiracies that would find place in the most fascist rhetoric of the early twentieth century. Our "education to reality" is both a scientific and theological or human enterprise. Anti-science groups easily confuse, jump domain boundaries, and are demonized or lionized. Right-wing anti-science is equated with both political libertarianism and conservative religions as preserving human values and innate freedom, while left-wing pro-science is equally equated with the same values, albeit with less divine origin of those values. And still, on the ground, they feel so very different. One is irrational and extreme while the other is rational and inclusive. Both, however, speak of the human in terms of freedom, human value as essential and derive from the history and failure of modernity. We are cultural zombies philosophically, theologically and scientifically—neither alive nor dead but still walking around.

And this is the point of this work—that we are in a transition point but that understanding the transition point is critical as to what science and metaphysics (for my part, namely religion and theology as its logical discourse of ritual and experience) is capable of, what they share in history and confusion and where we might end up. We are, as I paraphrase Kant, living in a scientific age but not in an age of science. This is, of course, not new. As early as the late nineteenth century, philosophers such as Nietzsche were writing about the end of the metaphysics of God, the early twentieth-century crisis theologians of the West were exploring what can be said about theology in a *wissenschaftliche* or reasoned world which included science as a discipline and method. In the late twentieth century, postmodern scholars felt that confusion was more normative than certainty. Culturally this is also reflected as we press for precise evidence-based pharmacology and better medical science (or even smart cars, etc.) while culturally advocating that naturopathy and osteopathy are to be admitted into healthcare plans and its practitioners called doctors. The local organic food shop has a mass of confused messages—take this natural elixir of herbal products learnt from spiritual traditions, which are shown in so-called scientific studies to ward of influenza/COVID-19 and yet are alternatives to big pharma/medicine and its medications that are unnatural and witches' brew of contraindications. How can this be? The postmodern or postfoundationalist would simply say, it is all one truth in a confused pot of herbal tea. Zombies?

This work is also highly personal. I have a parent who has Alzheimer's disease. Since the diagnosis and treatment of my father, our family has experienced the loss of my father as we knew him in pretty much every significant manner to us as a family. As the disease has progressed, he has retreated into a more historic self of his youth, and this erosion meant that much of his life lived as an agent or self is lost. He is increasingly a person limited by his brain's physical deterioration, which for him means that the present, the more recent past and so forth are lost and all that is remaining is some very young deep memory self. Most of his adult life, or memory of it, is gone for him in any way that is significant, from the people he loves to the things he did, be it as husband, father or even as professional. My father is his brain, at least in a way that I know him as Dad. And I fully expect there is a day coming in which he will not know his son. As a father myself, I often wonder at this truth of my father's disease and to be honest, fear it as that which I see myself above all else is as the parent to my two wonderful children. Like Phineas Gage, my father is not himself any longer, or at least how I knew him. Zombie?

The intention of this book is therefore a way to think about relationships. It is to ask about important questions—how scientific are we as a culture? Do we think scientific-like or do we still think in other ways about our world, and how do we navigate between those different maps of reality? I often note in the theology faculty that I work in that most future clergy or people who wish to work in faith-based organizations come from the humanities while most of those they serve or lead will, at least in the West, work in some kind of technological or scientific vocation and that this means the importance of understanding science and religion/theology is paramount as a parishioner might be a lab tech, engineer, technician, medical worker or educator who is working in the realm of microbes, computer circuits and scientific studies but comes to worship services to hear of a moral or social world. Just like my intrepid reporter, the clergy or faith leadership finds itself interpreting gaps without much understanding of science and what it is really saying, what it can say and how to live in that tension. This work is, however, living in the tension rather than arguing for a solution because of the very subtitle itself—Stone Age souls or ideas that come from a very different world of understanding the human person in the universe—physical and moral that are placed into modern minds. It rejects the idea of open conflict (primarily because I think the war is already won—we are not going to put away our cell phones and for most of us, given the choice between cancer medical treatment and prayer to treat cancer, we will always opt for scientific treatment) and the idea of integration of two worlds as a claim on a single reality. There is only one Real. The dilemma, I argue, is not which tool is the best per se but rather that humans cannot, by definition, claim the certainty of the Real that came with the old world of Judeo-Christian certainty (if this ever existed itself which is also untrue but a myth of modernity as modernity corrupted faith claims with science). It is better to understand the boundary of what can be said in both camps—science and religion—than to argue they are in opposition or in agreement. Both are human ways of speaking, and I explore this idea using the ideas of delusion or illusion to avoid being forever wandering zombies.

This work moves forward in a way I learnt by teaching, advising and supervising students over the years in both science and religion or theology. I am fully cognizant and more than sympathetic to the objection that religion and theology are not the same, but in the present context of Western academic and popular understanding of the words, I am using them somewhat interchangeably. For me, theology is the logical expression of a religious tradition's actualization and it is both messy and uniform. It, classically, includes orthopraxy or behavior as an outcome of orthodoxy (and this has become in many religious studies faculties the study of social science therein) and it is both easy to confound or confuse and to synthesize and tether them together. The broad trajectory is an explanation of the end of realism, whether metaphysical or scientific as humans have learnt that we are impossibly entwined as subject and object of our own minds and its extension to explore, order and understand the Real. Add to this, the confusion that what most of the West thinks is universal and normative as human or human self-understanding (the above values of individual and freedom) is in fact local and we move into a grave area of uncertainty. This uncertainty creates anxiety, anxiety creates fear and fear creates fundamental or foundational ontological statements on what is normative. The promise of Immanuel Kant's perpetual peace based on universal reason is as problematic as a normative Kingdom of God found in any Judeo-Christian tradition when one begins to pull at the threads of what is truly universal and normative. The reader will need to explore whether this ontotheological abuse—the intrusion of a universal absolute claim that in fact is local and contingent—is as much a problem as it has been. I would argue that it need not be if one is able to place a critical understanding of the Real into work, holding a humility in what can be said and focusing on that which is necessary for human prosperity, culturally, socially and morally. This, at least for me, is both a scientific and theological/philosophical task, but must be ordered as best we creators of illusions can to avoid being delusional. This work, then, is not an apology for an integration of science and theology as both expressions of the Real, giving both equal credit in conversations on the Real. It is also not an apology for the supremacy of science over faith or theology but does ask openly whether the persistence of Stone Age souls can remain in modern minds for much longer. And it worries what that replacement might be. It hopes for the best in what it might be.

Of course, there are a multiplicity of scholastic definitions on what is science, soul, consciousness and so forth. The reader might bristle that I am loose with definition; but, I will argue in response that the scholastic mode of strict definition belongs to misunderstanding rather than understanding as all those ideas—science, soul, consciousness, nature and so on—are heterogeneous in application and therefore in definition, except to buttress technical argument. Thus, famously, Karl Popper tried to define science in such a manner that few practicing scientists would be able to agree with as his narrow definition of falsification and empirical rigor. Popper is certainly correct that there is a tipping point in which scientists abandon theories in favor of new ones when the old theories are not sufficient explanations of phenomena, but this is not merely that one black swan means the statement of all white swans are false. The tipping point is linguistic. It is cultural and it is human and even intuitive. Science is a public method but it is also a human description of the world which at times slides into areas outside of measurement and theory. It is an "education to reality" but that reality is human. There is a transition in this work on how to think about science, but the astute reader will see that this was always in the bones of science as a method or strict definition. Likewise, "soul" and "consciousness" or "identity" seem interchangeable outside of a deep reading of the ideas for most people. To think of pious Grandmother in heaven is not to think of her except as we knew her, and few think things such as whether Grandmother's eyesight is fixed in heaven and so forth, what is her form there and what is the nature of her memory as construct and identity. In other words, we think of Grandmother as a person we knew and not as an abstract idea of soul. We are stuck with language, language coming from experience and then application of that language. For the most part, the definitions here come from how the ideas in the book run on the ground of language and application. Even Popper noted this, as he famously wrote, "Scientific discovery is impossible without faith in ideas of a purely speculative kind, and sometimes even quite hazy; a faith which completely unwarranted from the point of view of science, and which to that extent is metaphysical."[1]

Some readers might be further frustrated at times, especially those orienting toward either conflict or integration models of science and religion/theology. Those who see the foolishness of anti-science groups in fundamentalist or evangelical types, see the confusion of anti-science claims in the Spiritual but Not Religious types who advocate anti-vax doctrines, hold holistic medicine ideas while espousing the benefits of CBD on the so-called cannabinoid system and think that dinosaurs coexisted with humans might bristle that I am not advocating the abolishment of religion. Likewise, those who argue that there is indeed human truth, and even claims of scientific reality on the Real, in traditional faith claims might bristle that I am limiting those statements to the human order of understanding and ask that they be understood as such. Here religion and science work together as clergy function as counselors who use best psychological evidentiary praxis, understand that genomes dictate human sexuality and technology is welcomed as means to build communities, which is the goal of a divine order. It is possible, in that case, that there is a deeper spiritual reality, but I limit what can be claimed therein.

In terms of method, there are a couple of things to note. I try, for the most part, to argue within the logic of the two claims on the Real, using their own "logic" or expression. This means, for example, when I talk about neuroscience and religion, especially on the idea of the emergence of the soul, I use literature from those who are firmly in the camp that there is a soul and science gestures toward it rather than literature from neuroscience that would dismiss ensoulment outright as part of its "shut up and calculate" field of inquiry. Likewise, when I speak of the new physics and the quantum or quantum field world, I try to argue from the logic of the scientific rather than the metaphysical in the first instance—namely, what does it mean to them before what does it mean to the nonspecialist who then sees "pixies in the nanotubules" of human neurons and in multiverses that are places of divine action in M theory or 4D Gauss–Bonnet Gravity equations. Secondly, I also try to place the conversation into the context of language—after all, it is essential that both theology and science understand themselves as human language constructions of the Real—and as a result have a linguistic history that brings meaning as well as creates meaning. Our understanding of Nature is so poorly understood that we assume it entirely (pun intended) as natural, so that how humans construct themselves as a part of the Natural is critical

1 Quoted in Andrew Briggs and Roger Wagner, *The Penultimate Curiosity: How Science Swims in the Slipstream of Ultimate Questions* (Oxford: Oxford University Press, 2016), p. 434.

to understand. This is, however, not to say that no-thing can be said, but to understand more broadly. As a result, subsequent chapters are more about those ideas than a catalog of what science or religion can get correct or wrong. An illustration may be helpful.

Some years ago, I heard a very good paper on evolution and the 1925 American Scopes Monkey Trial from a rather good historian at a conference. What is clearly an epochal moment in some Christian rejection of Darwinism was argued to also be a political and social rejection of centralized education run by American Urban Eastern Metropoles (and their economics in publishing textbooks and extending mandatory educational age) by local rural communities and that this was as much in the water of the trial as was the science–faith conflict, which still is in foment today. Amongst the academy trained historians, that kind of argument is pretty evident and the complexity of that moment of American history is well travelled and understood, and yet popularly—whether in podcast, television or movie, news article or even sermon—it remains reduced to a simple conflict of Darwin and a particular Christian interpretation of the Bible. Both historical interpretations are legitimate. It is certainly legitimate to argue that the particular version of Biblical inerrancy largely associated with some American free Protestant churches and in particular its timeline of creation runs contrary to the claims of Darwin now understood as biological evolution, and empirical sciences found in geological, astrophysical and cosmological claims on large-scale evolution of all reality. There are clearly variations even within that particular biblical reading—a quick visit to a creationist museum to see a Flintstones's coexistence of dinosaurs and humans would shock those early creationists—but the sense that science, in all its form, is anti-Bible is correct for them. Likewise, the paper that argued it was the driver of urban–rural economic and social conflict into which the vehicle used was Darwin is also correct. That is why this work has its latter chapters devoted to trying to place the discussion of science and theology/religion into broader contexts of language, self-history and social construction. If you are looking for a strident new Atheist treatment or a strident apology for the supremacy of faith as intuitive and legitimate reason, then I am afraid this might not be, to paraphrase Ben Kenobi, "the book you are looking for."

The following work tries to make an argument of change not complete, and therefore not absolute. It is diagnostic on our zombie nature and offers a possible way forward. As the above example illustrates, it is also not merely largely two siblings—science and Protestantism eventually fomenting Evangelical or conservatives in the twentieth and early twenty-first centuries—but a whole series of complex issues economic, social and even political that frame the relationship between science and theology or religion. The conversation about science and theology or religion is a conversation placed into the human history of the West, and as colonialism occurred followed by economic expansion, it became a global issue. The conversation has to do with ranges of knowledge, the end or shifting of realism as a philosophical idea, the end or shifting of science from its largely Protestant roots and the rise of the industrial world, the modern West, the postmodern West and now the hypermodern world. It is not merely about rational domains of knowledge; it includes morality, economics and even emotions. It *is* the dissolution or death, primarily, of the idea that humans are supranature—long carried in the Judeo-Christian tradition as a bedrock idea—and therefore have a different relationship to the natural and Divine than animals, plants and the entire universe. Humans are not the central player in the grand cosmos, nor secondary characters caught up in a grand metabattle of cosmic divine forces of divinity rather just another ordinary thing in the universe. Being ordinary is difficult. Being ordinary, by definition, means one part of a much larger sequence of Being and therefore at the mercy of the parameters of that sequence. Being ordinary for the postmodern mind is being made and making at the same time. It is not the autocratic archangel of modernity and its freedoms. More sharply, we are in the world, universe. We are trapped to our own locations, and just as a bumblebee can only do what a bumblebee can do, so are humans located in our finite laws of Being. We must, therefore, "educate ourselves to reality." We have had a grand tool to do this—our notion of extraordinary self as souls—and this has allowed us to lift ourselves out of the sequence of the ordinary with nearly divine omniscient eyes to see patterns of divine will and order to which our best selves align. We were supranature. The history of science and religion is a clash of that idea, steadily being chipped away. But it is most imperative to understand this is not merely an argument on atoms and DNA, but it is also an argument on how we see ourselves as beings in the world. We are inveterate idolators—we cannot help but place ourselves in the center of the picture and make the drama about us. It is what we do, and it is profoundly theological. The question then becomes how can we understand this propensity? This work offers a partial solution after an accounting of the process, issues and places of both frisson and friction.

The argument flows as follows. Chapter 1 tries to understand what theology, read as the Judeo-Christian theistic tradition of religions, has to say about human beingness and where in general this rubs rudely against another narrative, derived certainly from that history, of science. In short, what we mean when we talk about "souls" and even why that idea was useful. Here we look at the theological idea of soulfulness. Chapter 2 tries to outline what exactly we mean when we talk about ourselves in terms of "minds" and the way in which the reduction of mind to brain (physicalism) is both understood as absolute by neuroscientists and subsequently configured by theologians and philosophers in light of that claim of neuroscience. What is argued is not so much the information (it seems the field is inconvertible) but how exactly theologians, philosophers and scholars of religion deal with the fact that we are, in fact, substantially that 3 lb glob of matter. Here we look at "minds" and the idea that the emergence of souls or minds is a nonphysical event, a spiritual coup d'état over matter. Chapter 3 is a response to the crypto-agenda of theologians (and frankly new-age charlatans) when they apply the new physics of the early twentieth century (quantum theory all the way to present theories of theoretical physics) as a realm of mystery and wonder. Here, an explanation of what can be said from those theories extending into theological or religious ideas is explored, what are the limits of said physics and why the marriage is odd for the scientist. Its surface attraction to the theologian is clear but I argue both a popular misunderstanding of the new physics and a fatal blow to theism. It is a cure worse than the disease. Chapter 3 begins to help us understand how we can educate ourselves to reality with our ideas of souls and mind as brain without retreating to the realm of mysterious wave and fields that seem like some Q continuum from Star Trek.

The remainder of the text works to help set boundaries and investigates the limits and uses of what kind of education to reality is possible. It thinks on what humans are when we think about ourselves, places us into a much larger cauldron of complexity than a simple soul that orients toward order or a mind that is capable of unlocking, without bias, all the universe's secrets. And of course, the sub-text of all this—that we might get past our tribalism that seems basic to religion. After all, our deity is the correct one—the butter must always be on top of the bread and this is divine fiat so that those who butter differently are antihuman.

What was the promise of early science and its corollary in some version of humanism clearly has not arrived—peaceful coexistence with each other and our natural world. It seems obvious that the erosion of a sacred canopy of ideas is terrifying, so Chapter 4 tries to place this conversion in a different manner than simply arguing that we must do one or the other. Instead, I try to argue how both visions of the human—soul and mind/brain—derive from an understanding of nature itself and that a kind of idolatry comes out as a result—one that is helpful for neither the advancement of science nor the engagement of theology or religion. By understanding the conflict or hoped for integration as starting with an idolatrous idea of absolute truth (having a divine fiat of truth therein), humans are not able to accurately speak or engage. This chapter tries to place us as emotional beings as well as truth seekers and argues that part of that experience of emotion is reconfigured when we drop our idea that we are supranature. Chapter 5 continues this thought and instead offers a vision of how one can hold ideas of nature, soul and/or mind/brain with an emerging confidence that avoids demonization and isolation in favor of a truer representation of reality, one in which humans participate. Here we encounter the idea that we cannot know the world in the manner either science or theology initially promised—as ironclad representation of reality. Instead, we are educating ourselves to reality and must avoid the pitfalls of mistaking our illusions, including the moral world, for the delusions of absolute reality. The penultimate chapter is how this participation might occur in the vacuum of the two polar opposites constantly argued in the book between science and religion or soul and mind/brain and avoiding (for me) the illogic of integration of science and religion as it is often argued as being two necessary aspects of a greater truth of Being. This includes thoughts on the nature of postmodernism, postfoundationalism and explores how we might work as a nature realist—a mind oriented or educated to reality—a mode of life together which is, at the same time, theological. Here I explore the idea of a nature realist, which is the place that I think we must conclude as the history of soul to mind. Nature realism ends our zombie existence. The end point of the work may be unsatisfactory for many as the erudite reader will note that while I think we still have a place for theology or religion to educate us to the Real, I am not entirely convinced that this is the final place, or even best, for human beings to understand themselves within our natural world, our universe and even our own social and moral worlds. We are educating ourselves to reality.

There are in any work many people to thank, for their patience, instruction, and generosity of time. I would especially like to thank Dr. Adrian Owen for his lunch time talks on neuroscience, and many, many others too numerous to mention as much of this work is from papers given over the last decade. Any misconceptions are entirely mine from my conversations with them. Finally, I dedicate this work to my two children: Gareth and Beatrice. Gareth is already on his way to a deep understanding of theoretical physics at the University of Waterloo and has a deeper humanity than most I know as well as a deeply scientific mind. Beatrice, while still in high school, is equally brilliant and her love of art, writing and science is inspirational to me. Both are intellectually curious, open to the world and generous human beings or nature realists. If they are indicative of the future of science and humanity, then we can look forward to a golden age of compassion and empathy. They are my truest joys.

Stratford, Ontario

Chapter 1

STONE AGE SOULS*

As a graduate student at the University of Oxford, I often explored medieval churches. Entering these buildings, and the twelfth- and thirteenth-century founded Christ Church Cathedral, I was always struck by how different the minds of that ancient population were to mine. Not so much that I lived in a society that used smartphones and computers, but rather that the very fabric of their universe and self-understanding was vastly different than mine. How they understood themselves with reference to the Divine, the cosmos and each other in strict hierarchy literally manifested on the walls and ceilings of the buildings. One aspect always struck me, making a deep impression on my psyche (as being a little chilling) was the presence of angels and demons or the supernatural in general. The world conceived in the Western medieval mind was that of us (humans), God and then another group of supranatural nonhuman beings. I was, of course, well familiar with these, not merely from popular culture, but more from my studies and readings of ancient Christian and Jewish theologians, whether Augustine, Aquinas or Maimonides right through to Martin Luther (and later). Each in turn had sections of theological reflection that treated the very real to them world of the divine other, not just in terms of God but in terms of another kind of being in contradistinction to human being. As a colleague once related to me of one of his mentors, John McIntyre, the premodern world had "us, Him (God) and them (supernatural

---

* I use the term Stone Age in its generally accepted use: As an archaeological description of hominid then human society which is 3.2 million years in the making and ending as late as 2000 BCE with humans moving to widespread metal tool use. The reason for the length of the period is interesting as it challenges one of the essential tenets of theological accounts of the uniqueness of humans, coming late in the story and caused, presumably, by divine interaction on natural evolutionary history. But importantly, it gestures to the fact that the brains we have come to have are the result of a very long selection by evolutionary processes and that the distinctions we make on human uniqueness, not only over other animals but indeed over other hominids, is evolutionarily speaking negligible. In short, we discover these ideas quite late in the hominid-human story. As we shall see, some argue this is evidence of divine selection and love. But this, again as we shall see, is a retro-conception (retro-con) that poses more threat to monotheism than relief.

beings)" as points of reference for human self-understanding. Humans were not only unique in this anthropology but unique in terms of two divine references. Much of what would describe human uniqueness—even familiar to contemporary minds—such a freedom, finitude and relation to God would be in contradistinction to the angelic and fallen angelic beings that were not free, not finite and in a different kind of relationship to the Creator and creation. In the description of human identity, there was a triangulation of identity. As one sees the process of modernity, this triangulation becomes a binary. By the end of the nineteenth century, the triangulation is binary—there is only humans and the Divine. The Divine is "humanity-plus" as projection or humanity is "God inside." And now, I maintain, there is only humans. We are left alone in creation, one animal in the vast chain of life. No longer unique and our idea of what might make us unique, imperishable or the like—the soul—seems to be problematized, or is it?

Much later, I would hold graduate courses on the doctrine of creation in Jewish and Christian thought, and being a course on beginnings, we started at the beginning with the Genesis accounts on the Torah and Christian Old Testament.[1] What the next 13 weeks revealed, moving through Patristic to medieval and into the early modern and contemporary accounts was just how the cultural questions of the theologian impacted how they understood answers on creation, cosmology, and anthropology in terms of Judaism and Christianity. Whether the earliest attempts by Jewish theologians to distinguish the monotheism of Judaism from the polytheistic ancient world or Pauline attempts to situate Gentile Christianity against Judaic concerns. This continues with Augustine's struggles against neo-Platonism/Manicheanism and Pelagianism, and slides into Aquinas's and Maimonides's attempt to engage with the Islamic version of Aristotle given by Averroes (Ibn Rashd). In our modern world extending to John Calvin's attention on election and divine sovereignty to answer the Anabaptists' evangelical faith while Schleiermacher's through to Karl Barth's attention on answering modernity's challenges in the realm of biblical studies, anthropology, psychology and science in general. Each, in turn, we discovered reinterpreted the seemingly solid traditional narrative (or definition) on the nature of the soul, freedom, immortality and therefore human beingness and relation to God via questions derived from the concerns of their period, from essentially new ideas on the nature of humanity

1 I use the term 'Old Testament' rather than Hebrew Bible to refer to the Christian Scripture's use of the Torah and Tanakh. I do so to distinguish it not only in form and organization, but also to recognize its different theological patterns of interpretation. The 'Old Testament' thus is the Christian reading of the Hebrew Torah and Tanakh. It is not, in my estimation, the Hebrew Bible and this alone belongs to the Torah and Tanakh.

and its relationship to soul, God and the host of other thoughts. Of course, this is not surprising as it is the role of theology to respond in creative fidelity to its time but what was surprising to the class (and me) was the nimbleness of the judgments in leaping over what seems to be, at first glance, immutable ideas in the Genesis narratives. Whenever our theologian looked at answering what is our soul (definition) and its relationship to God and others in creation (application), the grounds quickly shifted depending on the period we were engaged with. Contemporary ideas clashed with the ideas of the theological tradition, and the shifting was often radical. This got me to consider: Are we in an age of change yet again? How do we address the new vistas we see in terms of human consciousness, or our mind and our brain as they relate to our soul?

Of course, it is imperative to note that there is no one ironclad definition of soul or its application. Likewise, later, we will see the same on ideas of what is science, consciousness and identity.

However, there does seem to be a bedrock for these shifting ideas of soul. These shifting bedrock ideas in the Genesis text clustered on several points: (1) humans are uniquely created by God with a divine purpose; (2) humans are uniquely related to God on account of this creative action or Word of God; (3) humans are free to engage with this unique relationship, and in fact, some kind of afterlife or extended life is indeed dependent on this engagement; and (4) the account of divine activity is the evidence of the reality of the above points. Of course, there are a host of secondary ideas that result from the above account of the unique creation, relationship and destiny of human beings. These include (a) accounts on the nature of the soul and its relationship to the body; (b) accounts on the nature of the misuse of freedom (usually as sin or moral evil); (c) the role of faith traditions themselves in aiding the ascension of the self to its destiny (this naturally includes redemption); (d) accounts of the relationship of self/mind to body, especially in the modern period wherein psychological accounts conflate identity and mind. Finally, there are concerns that arise in any given period such as in the modern world the challenges of science such as evolution and psychology, challenges from the social sciences such as anthropology and sociology and finally challenges from the contemporary world on the nature of consciousness and human speciation in general. These last challenges are the focus of this present book.

However, it is important to note in the rather brief account on "beginnings," that what was thought to be bedrock ideas were rather clay footed in construction and in fact shifted frequently in the theological history. It is not enough to claim that humans have this kind of soul, or that kind of relationship to God, and therefore afterlife looks like this when we misuse our freedom. This is too simplistic, and the real history is far more nuanced, polyphonic and complex than simple reduction. After all, it takes but 300 years for the supranatural beings in Christ Church Cathedral to become curiosities seen in popular novels and media rather than as realities of human experience and reference. The world of "us, Him and them" is very far from us as we inhabit a world of us and the Other alone in which our interior life is the location of divine activity and connection. Just as we now refer to ourselves without reference to a "them," except as a presence of illogical or material insufficiency in terms of biology, we have new minds and souls as we dialogue with our contemporary world. What was once a triangulation between divinities is now, in the West, a binary in biology, psychology, anthropology or sociology, even in religious circles. Few clergies would consider, as did Luther or Augustine, that the human psyche is literally ridden by either a real God or the Devil in our choices or as Rashi consider the real presence of a good (*Yetzer HaTov*) and evil (*HaRa*) self as more than psychological categories or placeholders. Few clergies would consider original sin as an ontological category of being, and instead talk of it as misshaped freedom of the will. Even fewer, for example, would account mental illness to demonic possession! It is equally important, nonetheless, to note that the bedrock ideas above actually do form a general theological construction without which appeal to the supernatural origin and destiny of human beingness falls away completely. That is, the question to be explored is whether this host of ideas are really a Stone Age soul in modern minds.[2]

Why Does It Matter?

A quick troll though the internet on the role of religion shows deep polarities. Some New Atheists call for the abolishment of faith while many more pundits argue that interreligious dialogue is founded on a common belief in humanity and its humility within faith construction as a social good. Is faith benign unless radicalized, and hasn't it been largely an overwhelmingly positive force in the great cultures and traditions of the world? This question is an answer as well. For religiously inclined persons, this is certainly true whether there is a positive history that supports the statement or not. For religiously neutral persons, it is an acceptable statement only if there are proper boundaries on what domain religion can speak toward, represent and be socially conscious

---

2 This is an inversion of the evolutionary biology/psychology slogan 'we have stone age minds in our modern heads' in which the operating principle is that for most of human history (99% in fact) we have lived differently than in the recorded history of the biblical-Western culture and that this evolutionary history is the dominant determinant in human behavior quite before we conceived of souls, freedom and afterlives; let alone friendship or the like. See Cosmides, Leda, and John Tooby (1997). "The Modular Nature of Human Intelligence." In: *The Origin and Evolution of Intelligence* in Arnold Scheibel and William Schopf (eds), (Sudbury, MA: Jones and Bartlett) pp 71–101.

on. For religiously hostile persons, it is an untruth that neither history nor experience bears truth on. Most in the Christian historic West would probably fall into the first and second camps—happy to have religious voices with all the positive values that come with the traditions inherent in social and political systems. A deconstruction of the soul as an idea would seem to be academic or personal bias by the author. Surely there is nothing wrong with the idea, and indeed many of the great ideas of the Judeo-Christian and other religious traditions with voices of tolerance and hope. But is this true—not only in the historical sense (although it appears that the evidence of history is that religion has done a lot of harm unless you are inclined to the religiously powerful) but also in terms of moving the humans forward? Has the idea of a unique soul served its purpose, and what we need is a new idea that allows us to move forward as a species? The answer and suggestion throughout this work is "yes, it is time." The reasons will fall out more completely later in this work, but a summary will serve to mark the arguments. But there is also a caveat—we might still need the idea of a soul. Perhaps it does do a job or two in helping us as a species in that we cannot function as merely natural or animals. But let us start with the argument for its dismissal.

First, the idea of a unique human soul is hostile to the greatest threat to the human species in its history, namely climate change. The Judeo-Christian idea of soul carries within it an anthropocentricism that drives rampant disregard for other species and for the role of humans within the biosphere. This, in turn, has contributed to the denial and *laizze-faire* nature to which decisions on climate change (or other challenges raised by the use or abuse of scientific discoveries) impact humans. As humans disregard our own evolutionary prehistory, our place within the biosphere and consider the species to be unique because of its self-consciousness as creatures with souls, we will risk ourselves. Second, and related to the above, but even more pronounced in terms of population, humans will soon be far too numerous to sustain the species unless we come to new understandings of where humans can live, how we can power our machines and how we can feed ourselves without a massive population cull or technological intervention. The answers for these questions will not come out of the resources of Stone Age souls and cultures but rather by understanding our place in the universe, understanding our universe and its fundamental realities and by achievements that can only be described under the rubric of science. Our world is, and must continue to be, scientifically driven, and then an ethic or culture that builds on this understanding promises a future. Whether in court rooms debating the right to die, hospitals employing stem cell research, farmers utilizing GMO products to feed people or physicists understanding energy to fuel our world, the world to come is a world of science. It is shaped by the concepts of science, whether evolution or quantum mechanics or biochemistry, and the ethics of that future lags due to the luggage it carries from the Stone Age souls and minds of the great religious traditions. The brave new world we face is not bright, the only resource we have is not benign faith, even in ourselves, but in the knowledge that we can make decisions as we learn of that real material world. That is science.

As this work is prepared for publication, I sit in a room in the COVID-19 pandemic lockdown in Canada. Much of the above is brought to full light as to how perilous our fragile connection to nature is and how utterly powerless our vaunted sociopolitical, economic and religious systems are in comparison to nature itself. In four short months, a common virus shut down the globe, mocked all of the world's superpowers and gutted the stock markets, and the religious of the world retreated into agreement with science or openly mocked it as a hoax to control minds and power. Globally, we have consumed charts, listened to epidemiologists and medical officers and then read online all manner of articles while we all "caught" up our beliefs to the disbeliefs of going from freedom and life normal to lockdown. In some places, this was ordered and in others, chaos. On the eve of lockdown, the premier of my own province told Ontario residents to relax, enjoy travel on March break because our amazing health care system had them on their return, and two days later, the prime minister of Canada called for the return of said Canadians. The world changed in two short days for anyone not a scientist or medical person. Soon, we all spoke the "science of pandemic" or at least tried to do so without much understanding of science. And now we all hope in a vaccine, but in the meanwhile, doctors and policymakers are making horrible decisions on life and death and our vulnerable populations are exploited by the virus while the wealthy amongst us ride out watching streaming TV. Science, ethics, politics and economics are all on a collision course. Of course, in rapid fashion, Slavoj Ẑiẑek has written about this moment in time:

> The coronavirus epidemic can be seen as an assemblage of a (potentially) pathogenic viral mechanism, industrialized agriculture, fast global economic development, cultural habits, exploding international communication, and so on. The epidemic is a mixture in which natural, economic and cultural processes are inextricably bound together [...] "values and beliefs" should not be simply ignored: they play an important role and should be treated as a specific mode of assemblage. Religion is a complex texture of dogmas, institutions, social and individual practices, and intimate experiences where what is said and what remains unsaid is intertwined in often unexpected ways. Perhaps a full scientific proof that god exists would be of the greatest surprise for the believer her/himself. A similar complexity (or, rather, a rift) helps us to understand the

> belatedness of our reaction to the coronavirus spread—our knowledge was out of sync with our spontaneous beliefs.[3]

And:

> We are now forced to admit that modern science, in spite of all its hidden biases, is the predominant form of trans-cultural universality. The epidemic provides a welcome opportunity for science to assert itself in this role. Here, however, a new problem arises: in science also there is no big Other, no subject on which we can fully rely, who can be unequivocally presumed to know.[4]

The COVID-19 crisis, as argued by Ẑiẑek, brings much of the themes of this work together. That we are in an age of new "trans-cultural universality" and this language *is* science, and that this language is the only one that can reduce our panic (or not in terms of climate change, which is far more deadly in its manifestations)[5] as what we did to nature, nature is now doing to us.[6] As Ẑiẑek seems to argue, whatever is our new world order, science and its claims on reality, its limitations on the nature of ethics and its relationships to human behavior now seem a permanent part of our discourse. Or, as lectured to my students on the last week before we evacuated the university lecture halls for Zoom meetings, in moments of crisis, we look to the scientists of the world now and no longer the priests and politicians.

But isn't this merely another claim of scientific realism, loaded with materialism and a skepticism toward the mysterious which philosophers of science have warned us over the last century as toxic both to scientific discourse and to human betterment? Is this not merely dismissing an interior life, a connection toward each other and perhaps heading toward a totalitarian state of science with its *Frankenstein* or *Handmaiden's Tale* ghosts? Should science merely "shut up and calculate" and leave the morals and sociopolitical world to the religiously inclined (or in the case as suggested by Ẑiẑek, the sociopolitically inclined ideologist)? The danger, however, is not necessarily from the scientific realists but rather from the positivistic scientism-ists (if such a word exists). These, I charge, are those who confuse the claims of science as both material and true reality, rather than merely instrumental, with the claim that science acts as a kind of religion in itself—devoted to a method, dogma and commitment that is both

---

3 Slavoj Ẑiẑek, *Pandemic! Covid-19 Shakes the World* (London: Or Books, 2020), p. 117.
4 Ẑiẑek, *Pandemic!*, 125.
5 Ẑiẑek, *Pandemic!*, 69.
6 Ẑiẑek, *Pandemic!*, 81.

as complete, incoherent, and problematic as that which it wishes to replace in religion. Of course, much of the argument against scientism is its overreach in the natural sciences and subsequently into morality and ethics, largely understood as conditioned by the Judeo-Christian world and confused with Christian America or Europe as argued by MAGA redhats or Brexit supporters.

Positivism, Scientism and the Myth of Modernity

Many will object to the above on that positivistic scientism and its subsequent myth of science that promises a unity worldview of rationality over and against narrow theocratic or nationalistic bigot is false. John Grey, for example, is clear to articulate exactly this point—that science *qua philosophe* is neither a historical destiny nor a historical unifier. The past centuries—Soviet, Nazi and Western—have shown quite the opposite truth in that atheistic eugenics, dehumanization and belief in science as eschatological hope fly directly contrary to the historical record in which state sanctioned positivistic scientism has left the world less safe, sanctioned race and biological theories that have eradicated millions and justified political decisions that threatened the very survival of the species with nuclear and biological extinction. But as Grey rightly points out, science as savior comes from its historical emergence from Christianity, and the particularly Western form that emerged in the Latin tradition. Thus, it is natural to think of science as modern and messianic, particularly when charged with the left-Hegelian world emerging from D. F. Strauss and others.[7] Grey summarizes this as follows:

> History is a series of accidents; but if it has any discernable trend it is the growing power of human invention […] Positivism [as scientism] is a doctrine of redemption in the guise of a theory of history. The positivists inherited the Christian view of history […] they announced that by the use of technology humanity could make a new world. When they suggested that in the third and final stage of history [polytheism to theism to scientism] there would be no politics, only rational administration, they imagined they were being scientific; but the belief that science can enable humanity to transcend its human conflicts and create a universal civilization is not a product of empirical inquiry. It is a remnant of monotheism.[8]

7 John Grey, *Al Qaeda and What It Means to be Modern* (London: Faber and Faber, 2003), pp. 5–27.

8 Grey, *Modern*, 104–5.

It is confusing, however, to think that the thrust of Grey's criticism of positivistic scientism is toward theism per se; but rather it is an acknowledgment that theism itself and its heir—positivistic scientism—sought to make absolute statements on reality which then, by definition, exclude diversity and discord. It is because positivistic science grew out of Western Protestantism that it is infected with the same tendency to absolutize the Real. We, I suggest, need something beyond science at present, and that means moving to a more scientific understanding of humanity, which is largely metaphysically decentered and more natural in its realism. This is the final chapters of this work. But there is another gist in Grey and this is where science as positivistic scientism as a quasi-religion fits in. This is that religion is and was not *the* answer, apart from the fact that most theologies had an explanation as to why humans failed in achieving harmony and instead fell to the violent political and social of confused ontotheological abuse. This centrifugal part of the confusion inherent in the system had an accounting, original sin or living independent of the will of God in a nontheocratic system (Judaism and Islam) designed to limit this centrifugal force. In the Protestant world, this became an interior impulse, and thus both a social and psychological feature of anthropology. But all share the same concept—religion was a kind of imagining of human society that accounted, at some level, for the intrinsic inhumanity or animal within each and at a macrolevel. It did not eliminate the problem of violence, internal and external, but it did account for it or more precisely, describe it or name it. The scientistic-inclined positivists failed at this level according to Grey, and now its peculiar optimism stands nakedly amazed as technologically informed fundamentalists reject them and the entire modern project. In the end, it is theism and its particular form within the West that caused its own isolation and humanity could not remain naturally natural and so it has return to the metaphysical and the fundamental (see Chapter 4).

But does this mean, as Grey also argues, that science is merely accumulation of knowledge that is inert until it is married to some kind of unifying fiction, which we call politics or human life as culture?[9] It is here that it can be, as in Chapter 5, an "education to reality," which includes the possibility of it being a myth (e.g., not impartial human knowledge) and which can equally gesture to the lack of performance by self-conscious human beings as to any uniformed ethical reality when it treats that expanding knowledge as the best tool to criticize and establish human myth making as civilizations, even in conflict? Grey admits that "myths are not refuted. They simply disappear, as the ways of life from which they sprang from fade from the world."[10] And that

9 Grey, *Modern*, 110–11.

10 Grey, *Modern*, 118.

in terms of the Western cultures, this is also a belief in scientism that leads to secularism, enlightened and peace, which is contrary to its own reality as well as objections by others who believe that terror in the name of religion can lead to a peaceable world. But what Grey does not admit is that science is not scientism, and that it has not quite yet been able to overcome the most persistent belief undergirding the monotheistic traditions—namely, the belief of the uniqueness of the human person and the human soul as especial from all other things in the physical real world. This foundation, contrary to Grey, is the problem that science has not overcome. It is not that humanity has been abstracted by science to a series of random genes, but rather that there is a belief that these series of genes somehow make human culture special, above the laws of the universe, nature, ecosystem and so forth. It is that we do not, as Grey would later write, admit to our animal nature, or I argue our nature realist position. Once we have agreed to this fiction, we can escape the delusion of the modern theist, scientist and the postmodern postsecular fundamentalist or terrorist. It is also the beginning of the solution of the real problems the species faces, not the intermittent problem of history—namely extinction due to climate change, population or fouling our ecosphere. This last point, Grey intimates when he writes as the conclusion of his essay on why we do not know what it means to be modern: "Yet it is the interaction of expanding scientific knowledge with unchanging human needs that will determine the future of the species."[11] Where he notes this in terms of overpopulation and mass destruction as politics, it should be pushed to the conceptual as well. One unchanging need of humans is to ask questions of existence and then to frame knowledge of the world accordingly. If we learn to ask questions to reality more clearly, more universally while acknowledging that need to create fiction, it is more likely that we will walk away from the brink of destruction as the reason we are facing destruction is the relic of our old ways of answering unchanging human needs.

Returning the Bedrock of Essential Concepts: Do They Theologically Matter?

Trying to describe a normative Christianity (or Judaism) is like trying to describe a normative human being. Just as humans have 23 pairs of chromosomes, four base pairs of 20 amino acids that organize as some 3 billion possibilities, so are the possible forms of religion, even familial as Christian or Jewish, seemingly infinite. There exists a continuum of possible configurations in both history and

---

11 Grey, *Modern*, 118.

in any singular moment that make it impossible to decry, despite the attempts by theologians and traditions otherwise, that this is the final and definitive manifestation. But, as in evolutionary biology, we can use cladistic analysis to group a family of religious tradition in relationship to an originating concept. That is, we can describe a set of originating ideas that interrelate to define Christianity or Judaism as belonging to one family or another, and it is those ideas that theologically matter. If they are lost, or substantially changed, then there is substantial alteration of one key foundation so that the resultant branch (clade) is in fact something new. The easiest example of this is clearly the break of Christianity from Judaism, in which although they both share much theology and therefore have a common source, the claim of Jesus of Nazareth as the messiah is clearly a different theological branch than in Judaism, both in its time of continuity with nascent Christianity and in the rabbinic form that came to be normative post 70 CE. What is proposed following are key theological ideas from which any substantial deviation creates in fact a new branch and therefore new interpretation from the normative or family tree. These will work to speak of the "soul" as the religious aspect of the human being.

Before launching into a description of the theological family of the Judeo-Christian tradition, there must be a moment to articulate the presumption of normativity of Evangelicalism and Fundamentalism in the larger North American but also through the nineteenth- and twentieth-century missionary movements into the global world as biblically correct. Trying to argue why this idea is false is in itself an entire work and there are volumes, even within its own publishers, that draw attention to both the history and differences within the movement that most associate with Protestant derived Evangelicalism or Fundamentalism.[12] Certainly, given both the numerical strength of these loosely grouped movements in terms of members who identify as Christian and certainly political or social strength of them, they are rather important to understand as movements in themselves. But this, at the same time, is not to assert they are either historically or contemporarily "normative" Christian theology or worldviews. To do this would be as if one were to argue that there was a normative medieval Christianity, normative Protestant reformer or even a normative early Church community when what we do have is a family of similar clustered ideas, or variations on themes. It is similar for Judaism, Islam and other world religions and there are lots of scholarly works that help us understand the presumption of normativity of one strand while the reality of a complexity is apparent. We, as I often will argue, are inveterate idolators and this means simple classifications or a heurism.

12 See, for example, my major coedited work: Alister E. McGrath and Darren C. Marks, *The Blackwell Companion to Protestantism* (Oxford: Blackwell, 2004).

Broadly speaking, there are four single points from which Jewish and Christian theological self-understanding derive: (1) humans are uniquely created by God with a divine purpose; (2) humans are uniquely related to God on account of this creative action or Word of God; (3) humans are free to engage with this unique relationship, and in fact, some kind of afterlife or extended life is indeed dependent on this engagement; and (4) the account of divine activity—whether Torah or Jesus Christ—is the evidence of the reality of the above points. A further group of secondary ideas seem to follow but these do not seem to be intrinsic or foundational (despite creedal affirmations) that include the unity of God, the presence of sin or some other account of human evil and some account of the mechanism from which this relationship or election by God to humanity works. The reason I do not account them as intrinsic is that these seem to shift over the history and struggles theologically within both traditions.

There is one notable omission that may, on the surface seem incontrovertible, and this is the idea of monotheism, which is often given as the *sin qua non* in the Abrahamic traditions. Monotheism, however, does not seem to be of the same level as the above if one reads the accounts of both Judaism and Christianity. The Torah is clear that there are other gods—as the first commandment and the *Shema* makes no sense otherwise if there is one God and a host of other "less than" beings. The *Shema* is a claim of singular purpose of Israel's God, not of the ontological rarity. No one or Jew can read the Torah otherwise. Likewise, later Christian claim of the Trinity—a concept staggeringly wrong for Jews and Muslims—is a statement of unity of purpose situated in a mystery of linguistic confusion that no Christian theologian or creed has ever been able to parse. Later angelology does indeed populate the cosmos with divine beings, but the simple fact is that the Christian claim of God in Flesh (particularly in the person of Jesus the Nazarene and then in the ubiquity of person of Jesus Christ the Ascended in the Eucharist) seems to all other monotheists impossible. Monotheism, it seems, comes out of a Greek metaphysic in which the singularity of God is adjunct to purpose due to a doctrine of divine simplicity or aseity that perhaps could never carry the burden of the idea in both traditions. This is not at all in the biblical tradition, and historically has rested uneasy in it. The sheer theological history of Christianity—whether the Arians or the Eucharistic controversies of the sixteenth century—bears witness to this issue of definition of strict monotheism with divine simplicity or aseity. Yet there is yet another layer of difficulty in this reduction of the importance of monotheism in that rejection of its defining phylogenetic source is not the same as accepting pantheism or God as a force or energy in the universe. Whatever Jews and Christians mean in their self-understanding of God, and the relationship of that God in the possible pantheons of origin or cosmology, what they do mean is that this God is personal, historical and interactive with the people of God. Everything hinges on this point—God is a being with whom we have a present and future relationship, and the life circumscribed by adherence to Torah or Church/Christ is a life of interaction with that source.

Thus, we come to the theological constructions of those fundamental ideas.

Humans are Uniquely Created by God with a Divine Purpose

Without doubt, one of the leitmotifs of the Judeo-Christian story is that humans, and this historic group of humans in either covenant, are particularly uniquely bound to God. This claim is seen in two engines of thought that both traditions run along the tracks of communal practical life. The two ideas are creation and creation out of nothing (*creatio ex nihil*) which comes later in the Christian tradition, and which is largely adopted theologically in the Jewish traditions.

The idea of creation is an idea of intention by the Creator. In contradistinction to other ancient mythologies that proto-Judaism sprang from, there was no accidental or incidental creation of the world in which humans inhabit by accident, happy or otherwise (and the stories are often borne out of violence). More to the point, humans themselves were specifically created (and loved, set apart, made holy, etc.) by God. Unlike other mythologies, humans are placed in the center of the drama of creation, so much so that the entire history of creation is really a history of the impact or influence of humans on creation itself. The theologic is clear in the Genesis accounts: God generally creates the world and specifically creates humans within it; God gives humans a privileged position (usually in the naming of creation that denotes ownership) within the economy of creation and which distinguishes humans from other creatures; humans behave in such a way that the foundational parameters of the initial creation is forever altered (how powerful we are!); and finally, this alteration of initial parameters manifests further divine intervention in order to restore or renew humans to the original set of parameters. Careful theologically trained readers will, by now, be screaming objections to the language of the above. The use of "initial parameters" will seem offensive as it is theologically vapid; surely the cycle of the fall of Adam and Eve, the Flood, the call of Abraham, Sinai covenant and final coming of Jesus of Nazareth is more than an alteration of parameters?

But that is the exact point: it is not clear what those parameters were within the traditions—both Jewish and Christian. There are a multitude of possibilities as to what exactly was the penalty of the disobedience (or even if it was indeed such in the Rabbinic traditions) and what exactly constituted the initial created purposed state of humans in Eden. Each event that constitutes continuation (or indeed supercession) of the previous divine intervention to human influence in the divine plan for creation and humans in that creation has been interpreted differently within the traditions. Sin—human intervention in creation which impacts ecology to society – is clearly foreign to the original plan of Creation and required divine reparation or renovation to restore creation and humans to their intended place with God. How this is done is understood differently across time, tradition and audience. If we cannot find a definitive answer, then we must look globally at the intentions of the biblical stories and subsequent theological accounts. Thus, we are left with a strange set of concepts that rest on the simplest account—creation is purposed, and humans uniquely influence the state of creation. With the arrow of time, humans interfere while God responds, and a dialectic is found in this real-time history. Of course, what is the nature of that purpose account naturally follows and the answer, again, is rendered most simply that this purpose will be found in the subsequent activity of God in the history of the people of God. Theologically astute readers will recognize this as the doctrine of election (e.g., that only a specific group is chosen by God to participate in this dialectic of divine–human activity), and we will return to this idea later. But presently, it is enough to deduce that creation is purposed, and this purpose will be revealed (literally) in subsequent divine actions within the history of the people of God.

We then have a series of foundational ideas common to both traditions of which the most important include the uniqueness of humans in both the created order and to a particular set of humans elected by God to accomplish divine intentions within the creative act.

Humans Are Uniquely Related to God on Account of This Creative Action or Word of God

Given the centrality of humans in the Genesis story and the subsequent involvement of humans in the ongoing arrow of divine time, it follows there must be an account of what mechanism makes humans unique to other creatures, and then, by extension, what makes this historic group of humans uniquely suitable for involvement in divine schemes. The mechanism is given quite early in the biblical stories—humans above all other kinds of creatures seem to possess a "soul," which is later given as the "image of God." At its most basic reading (and again, there has been much ink spilt on the meanings of the various Hebrew words in the Genesis accounts), *human* souls allow the potential for interaction with God. In even more simple terms, theologically this means humans have unique properties due to the creative Word of God that allows humans to interact with the divine will and purposes. In the majority, this is uniquely and exclusively a property of human beings. Other creatures have life, and perhaps even a kind of soul, but only humans have *this* kind of soul that allows interaction with God.[13]

Later theological accounts will give shape to the nature of this soul property, locating it in various places such as mind/intellect, emotion (for ethics), consciousness, God-consciousness, openness to Being (and being) and so forth, but the barest concept is that only humans have this unique relationship and by extrapolation that the decided attribute of "soul" is part of the Divine itself. The broadest way of speaking of this is that humans share in some divine imaging of God. For some more contemporary theologies, God has some human imaging as a part of divine process, so we returned the favor.

Of course, once more the theologically astute will object to this point there has been no mention of the body or finite capacity of human beings. Certainly, the difference between Creator and creature is not only of degree but also of kind. Creatures, including human creatures, are in the biblical tradition both of Creator and of the material world. Here the concept of embodiment is introduced as an idea. Humans, both in the accounts and by experience, are certainly material. Thus, to be created is to have the properties of a body and a soul, and this is not accidental to human beingness but essential. But to consider this aspect, human uniqueness is to be ensouled leads us to the next idea in the two traditions we are thinking alongside.

Humans Are Free to Engage with This Unique Relationship, and in Fact, Some Kind of Afterlife or Extended Life Is Indeed Dependent on This Engagement

If being human is to be ensouled, or the proper state of human beingness is to be material and spiritually related, then it follows that an account of afterlife or extended life must follow. The alternative within the traditions has been baldly rejected (although there have been adherents) if this is not the case. In this case, human creation is but an intermediate state—a cosmic weighing station—in which the accidental (and seemingly divinely ignorantly purposed) coming to existence of this kind of beingness is merely a stage transitioning from the original to the original state of spiritual perfection. In more simple terms, as in the neo-Platonic traditions, material existence is an unfortunate stage in a perhaps endless cycle of perfection to imperfection

13 For a thoroughly entertaining and accurate account of some of the infelicities of conception on the human soul in our both near and far common intellectual history, see Mary Roach, *Spook: Science Tackles the Afterlife* (New York: Haddon, 2005), pp. 57–75.

to perfection. This has several theologically unfortunate problems for the theistic traditions.

First, it makes God rather hapless, perhaps cruel, and certainly not powerful. There is something, let us call it the universe, which is more powerful than this God. It can be argued that this God is even less powerful than the creation itself, which once it is running works to its own end independent of the Creator's purposes. For the theistic traditions, such considerations make the preexistence of human souls (and hence the idea eventually of creation out of nothing or *ex nihilo*) and the future return to that preexistence very unpalatable. Whatever this creation is, it is certainly not accidental so that embodiment is essential as a property of human beingness.

Second, it makes the material world itself rather hapless and problematizes most of human experience that comes out of the body which, by extension, is also human society. Not only does it privilege the spiritual or immaterial (which is a problem recurrent in Jewish and Christian theology) but it also makes human society problematic unless it is some kind of isolated monastic driven isolation (and here we see a source for such encratic existences) and that seems on the surface to be different from the point of bringing about divine purpose by the people of God. Certainly, in both Christian and Jewish mystical traditions there is a belief that some holy men in their piety bring about a secret movement of divine purpose as a kind of spiritual illuminati, but this seems to be contrary to the plain sense of the biblical and theological traditions. God's entire people, not just the holy few who practice x-piety, are involved in the purposes of God. Special (and even secret) election of people out of all humanity seems to be an all too human practice, and one that seems to read against the accounts of creation. This is certainly the origins of priesthood in the Jewish tradition and cannot but lead to the thought that this seems to be restrictive and ontotheological rather than as intended.

The reason for this consideration is simple and takes us to our final concepts. Recounting a story in which humans are involved in bringing about the purposes of creation in which bodies matter means that creation matters (point one above) and that this people of God must have the ability to participate or not in this purpose (point two above). Considering a special largely encratic elite whose goal is flight from material life breaks on the cliffs of both basic ideas—creation is purposed, human creation is likewise purposed, and therefore the state of human creation matters—not only in the beginning but also in the end point of the Creator. The language that is used to explore this idea is the afterlife or extended life. Both ideas, then, find themselves in currency in both traditions.

If "matter" matters, and human beingness is the proper amalgam of spiritual and material existence or soul and body, then the question of what happens to that body becomes paramount. Later traditions would engage more deeply on this consideration, and whether there is dualism, physicalism, and so forth, but for now we are merely looking at the theologic of the bedrock concepts. There seems to be but two possibilities when considering soul and body—one is afterlife (and its correlation of nonexistence outright) and the other is extension of life as to maximize the form before cessation of existence. Thus, the two broad positions found in the traditions come to a consideration of an afterlife and an extended life. Both can be, and have been, read in the Genesis accounts and in the traditions in a host of possibilities from extraordinarily long-life spans to general bodily resurrection to live another extended life, to a new kind of body that lives in eternal bliss to simple nonexistence. What all share is a fundamental conviction that creation's unique purpose for humans is to live as the soul–body modality as the proper form of human existence and that apparent cessation of life is not an obstacle to this. One has a new chance or merely enhanced enjoyment.

This gestures to the final concept to be considered—if the proper form of human existence is to fulfill divine purpose as soul–body creatures, there is some enhanced life resultant of living in that awareness or relationship, then why don't we all do so? The empirical reality of theological judgments is that humans do not, in fact, do this at all. The answer as to why is the theological account of sin or freedom.

Sin as an empirical reality (as my old supervisor commented) in both traditions is the reason for living in the divine society (Torah or Church) accomplished by divine will (election in Torah or Christ) and the excluding condition for why humans are not de facto already living in the privileged end point of creation as the afterlife or extended life. Certainly, there are accounts of first falls, pre-falls and nonreality of sin in the tradition(s) but these are well outside the theologic of both traditions. The reason usually given is blindingly simple, despite being deeply theologically nuanced, and this is that humans are given the choice to fulfill their spiritual destinies. Eventually, this would become an account of free will, but it does not need to be one. It can merely be a description of human materiality (our bodies are the reason we are not free) or lack of election in contrast to divine strength as in John Calvin (the elect elected before any action, and independent of action to demonstrate God's blinding power). Regardless of the nuance, whether free will is ultimately a theological chimera or true, there is a foundational affirmation in the idea of sin that humans do not choose to align themselves with their divine destiny, which is communion with God. As a result, they endure the hardship of that failure—both in this life and in any afterlife. In other words, this life ethically matters just as "matter" matters. The logical circle is complete.

Human beings are free (and again, there are different accounts as to what exactly that freedom means) to fulfill divine purposes and their participation in divine will manifests an account of extension or afterlife of this soul–body complex called a person.

The Account of Divine Activity—Whether Torah or Jesus Christ—Is the Evidence of the Reality of the Above Points

The final consideration is most problematic for modern minds, especially modern minds that consider the self as an isolated psyche homunculus floating above the interrelated web of social relationships that underbuild a person. While this is the gift of modernity in many ways to conceive of individuals rather than societies, and this has given us much, it is also the case that it is patently not true. This individual soul is not the entirety of creation, despite the philosopher's claims otherwise. Instead, what the theologic manifests is a much more robust idea, and the root of the problem—that group is more important than the individual because the group is alone the sum of the divine will and divine purpose. It is blindingly simple, but also incorrect, to read the narratives and follow the logic that the one individual soul matters (and even that this is taught by the rabbis and Jesus of Nazareth) and that God's love is presumably so compelling that God would have done all of creation for but one person. This is a nice thought, but it is not at all the narrative or the theologic of the narrative. The fact is that the narratives, including individual souls, are placed into a much larger account of many souls, so many that God elects some small group (Jews and then Christians) to bring the wider will of God to the rest. The theologic is remarkably social or collective, while individuals make choices within, the narrative thrust is collective. The circle is completed when the people of God manifest evidence (*Tikkun Olam* or parousia of Jesus) that divine will has been completed. Participation by nation or Church is evidence of the initial creative Will and Word of God. The means by which participation is measured is group conformity to Torah or the Church between the tick and tock of material time. Thus, the proof of the logic of all the above is found in the traditions of the Torah and the Church as living evidence of divine concern or purpose or presence. God is not without in creation per se (although it may be theologically true to affirm this) but rather God is in the community of God. This is the place in which the community must turn to for understanding of the theologic of the ideas of what it means to be created, ensouled creatures. But, and crucially, it also provides a set of boundary conditions from which to consider those possibilities and the entire history of both traditions is a continual reading of that narrative within the constraints of above.

It is not that there have not been alternatives to the above basic considerations and that those have not met with often brutal, bloody and intolerant opposition. Rather it is that the above ideas root the tree of Jewish and Christian theology and developed the idea of the soul as a manifestation of the internal logic of ethical human life together, warranted by God's divine purposes. Every other theological claim it seems to this author runs along these basic tracks of thought shared between the two traditions and from which all present inquiries as to how the twenty-first-century mind can understand the idea of the soul as human beingness must engage. It is completely possible to start a new project based on new understandings of the human person and therefore soul, and in fact, this work proposes just such a project. One could easily place the metaphysics of cognitive science here and introduce ideas of consciousness and identity/agency as found in experiment and social behavior. But it is also the case that it is time to measure the utility of those Judeo-Christian theological ideas and ask whether morphing them to fit with present understandings of the body as gleaned from neuroscience and evolutionary biology can be accommodated. If not, then the ideas must and should be left behind as belonging to a world mind that no longer exists except as a faint echo or present ghost. This is our present task—to ask whether this Stone Age soul can exist in modern minds.

Chapter 2

MODERN MINDS—EDUCATION TO REALITY?*

The question of this chapter is simple: what do we mean when we speak of ourselves? Is to have a mind, be a self and be conscious the same thing as to have a soul? And what of our bodies and particularly of our brains which seem to all moderns evidently the vessel of the soul/mind/self and consciousness? But before we look to this, we need to ask whether when we are doing theology and doing empirical (or even the human) science we are doing in fact the same thing? We start here because if what we are talking about is really two methods, then confusing them is patently unwise. But are we confusing two methods?

Origins, and Models of Science and Theology

While many theologians involved in the science–religion/theology debate work to distinguish and validate the processes of method between the two magisteria and argue for appropriate discourse for each claim on the nature of reality or offer criticism on scientism's overuse of its own heuristic method, most fail to engage in a critical analysis of what is really at stake. What is at stake is the tacit driver of all questions, reactions, rejections, détentes and combinations for the simple and historically obvious reason that the development of science and by extension our modern minds as an offshoot, and its subsequent philosophical assumptions, is only possible in a Judeo-Christian milieu and that its development is symbiotic, or worse, parasitic and supplants the very worldview that permitted its birth. This is, of course, the logoscentrism of the Judeo-Christian world or more succinctly the belief that an ordered, lawful universe exists outside the experiences or minds of human observers and that, in part, our *imago dei*, is attuned to this greater reality. This logoscentric perspective is the concentration of the four theological bedrock

---

* Sigmund Freud, (1928, 2010), trans. W Robson-Scott, *The Future of an Illusion* (Mansfield: Martino) p XXX.

ideas in Chapter 1. But there are two glaring facts that any such consideration on the value of this idea must face.

The first is the historical fact that the modern scientific worldview that permeates our lives from smart phones to combustion engines to pharmaceuticals has indeed altered our self-understanding. We have modern minds, and some of those categories of thought in key areas such as identity, sexuality, behavior and even sociality are substantially altered by our understanding of the science that underpins the area of focus and the use of technology that accompanies such understandings. The second fact to be considered is that those new mind maps may be, in fact, contrary to the original impulse or starting point of our modern world. The question, briefly, is how we came to be at this juncture with our Stone Age souls in modern minds?

This brief history is best summed by the affirmation that our nature is to understand Nature and this is part of divine plan for the religious mind. In other words, we should not be afraid of science when it is in dialogue or integrated in our faith theologies. I am reminded of Robert Jastrow's often cited misquote that the scientist works a lifetime devoted to reason and understanding only to scale a mountaintop on which the theologian already sits. That claim is, in part, based on a unique history, and a failure of the Enlightenment to deliver its promise. As theologian Eberhardt Jüngel summed as the Enlightenment's perpetual failure, humanity could not remain naturally natural and ethical in that naturalness. For Jüngel, this was a defense of theology and revelation in a world devoid of the supernatural and having lived through the horror of the twentieth century's atheism and scientism,[1] but it can also be understood to support the idea that it seems part of human nature itself to posit itself as being special, above (supra) Nature and therefore unique. Of course, as philosophers of science and theologians point out, this is also a cipher as the very idea of Nature itself is a construct in our postmodern sense as we cannot lift ourselves out of our very context as humans in nature. But, as historians of science repeatedly argue, this was not always the case and a simple confidence on the relationship between the human and the natural greater world—classical realism—is largely a gift of the Christian Platonic worldview that has preoccupied the great Church traditions since the third century CE. What was once assumed—humans had a privileged place in, and could understand, Nature was both necessary for science but now largely rejected for a form of critical realism, which places the human mind into nature.[2] Add to this, the

1 Eberhardt Jüngel, *Gottes als Geheimnis der Welt* (Tübingen: Mohr, 1977).

2 See, for example, the excellent work of Alister McGrath, *A Scientific Theology: Nature* (Grand Rapids, MI: Eerdmans, 2001).

transition of the idea of "soul" to that of "mind" followed by "person" and "brain" in the last four centuries, and one has a rapidly moving target being defined.[3] And so much of the integration, dialogue, criticism or rejection between the potentates historically have focused on discussions as to limits, parallels and spheres of competency to discuss the Real, and subsequently on the nature of the Real as knowable, which includes revelation as a viable mode of knowing.

The critical realist school of various philosophers and theologians shares the work of correlational giants of theology such as Paul Tillich, Wolfhart Pannenberg and Edward Farley. All are essentially trying to find a way for revelation—whether biblical, Eucharistic or ecclesial or in terms of religious experience—and find a stake in defining the Real that largely empirical sciences claimed to describe. In these wide swaths, what we see in the conversation is a discussion on the multiplicity of reality, modes of discussing them with specific competencies and subsequent ethical and humane outcomes. The physicist, biologist, psychologist and theologian, as such, all discuss the nature of human reality but do it with different but similar end orientated results, which is a cosmic humaneness. Of course, the endpoint dictates the more substantial place of theology as it deals with the ethical, whereas, and rightly so, science is a method of inquiry. But, for now, this eludes the formative aspect of having a "modern mind" informed by scientific understandings, which is not, by definition, a/ethical.

The most common trope is that both science and theology work in different spheres of competency, both of which work together to form the most complete picture or understanding of the Real and that conflict comes about when the two magisteria confuse whose role is which. This common defense—wisdom versus technical knowledge—is the one issue that cannot be argued away or simply constrained to a boundary condition, and this is the specter of the entire Platonic tradition within the Great Church traditions as we have come to understand them. This is the belief in a kind of ultimate unity of knowledge because the source of understanding is One.[4] Whether understood as a kind of perspectivalism, as a kind of relativism of knowledge to its degree of explanatory use or as a kind of postmodern agnosticism, there

---

3 On this well-traveled group, see the transitions from "soul-talk" to "mind-talk" in Malcolm Jeeves, "Mind Reading and Soul Searching in the Twenty-First Century" in *What About the Soul: Neuroscience and Christian Anthropology*, ed. Joel B Green (Nashville, TN: Abingdon, 2004), pp. 14–15.

4 This is assumed in most theological-scientific works as best described by John Polkinghorne that the "universe [is] shot through with sings of mind" and that this mind is the unity of God. John Polkinghorne, *Quantum Physics and Theology* (New Haven, CT: Yale University Press, 2007). p. 8.

is the tacit belief that there is a Grand Unified Theory (GUT), but that humans are unable to fully describe or understand this Unity.[5] Of course, in this GUT is the additional data of the Scripture or revelation, whether text, incarnation or ecclesiology as religious community. Thus, as A. E. McGrath noted, the classical two modes of knowing God—via reason in the natural world and through faith in revelation, both point to the same Unity.[6] I use that term—GUT—purposely, mainly because I want to argue that it is a misunderstanding of the poststandard model of physics. Too often, the seeming crack in the surety of the standard fixed model of the universe as completely rational and ordered is replaced by a cacophony of chance, multiverses, multiple states of reality which is attractive to the mystically minded. Once the sterling champion of the Real—physics—was demonstrated to be far more mysterious than imagined, the gate of the supranatural is opened and the world is suddenly far more agnostic than before. Foundationalism, whether theological, metaphysical, or even physical such as in Carl Hempel's defense of "Inference to Best Explanation" requiring one constant, is rejected. But, and this is an important point, is this *really* what is being argued? Has the quantum, the complexity of systems therein as being theoretically nonreducible and therefore allowing for multiple understandings simultaneously, opened the door to the mysterious and revelation?

This, I think, is the goal of scientific method to sharpen to this point, and this is the point at which there is visceral reaction from the religious community because it senses that this asymptote is a steady march against theological understandings. Thus, the early anti-science theologians focused on the issue of analogy as the only mode of knowing, recognized that the claims of miracles are problematized, the claims of human–divine concord and particularly in terms of divine personalism is challenged by Darwinism, and that modern astrophysics and cosmology strike at the center of a belief of anthropocentrism. In short order, the objections on the social impact, as orthodoxy drives orthopraxy, come quickly as there is a general worry on the morality of such metaphysics, or the lack thereof.[7] There has always, then, been the seed of destruction within the two siblings, and the short history of which is that the science-sibling cannot be restricted merely to technical understandings. One does not have to look very far to see how this is a flashpoint—whether Darwinism as a road that leads to atheism in the American

5 Polkinghorne, *Quantum Physics and Theology*, 2–3.

6 A. E. McGrath, *A Scientific Theology*: *Nature* (Grand Rapids, MI: Eerdmans, 2001).

7 Polkinghorne notes this very sharply when he writes: "I am entirely convinced of the existence of quarks and gluons, but that belief, illuminating though it is in the limited sphere of elementary particle physics, does not affect my life in any significant way outside the pursuit of intellectual satisfaction in the study of the laboratory. In contrast,

Fundamentalist movements or the opposition to Young Earth Creationism by paleontology and geology. Of course, these are extremes and many of the specifically Christian traditions have very little truck with such ideas today, but this ignores the fact that almost all Christian traditions initially had some level of opposition to Darwin.[8] It was theology that moved, not the sciences themselves. And the movement, I suggest, was a movement to the interior as a location of religious truth. This, then, meant the soul. The challenge of science pushed theology to defend the soul as the one true location of divine action or difference or intention. The rest was left to science. The flashpoint I wish to explore in this work is the relationship between neuroscience and theology and the merely physicalist position that all we are as human beings are brain.

Let's Get Physical

The problem in terms of neuroscience and theology is that it is increasingly evident that personhood—however defined—is a function of the physical properties of the human brain whether simple or complex interactions. Our "self"—again a terribly loaded term—seems to be related to our physical brains whether in appetites, reason, or even spiritual inclinations. This has been shown over and over by studies largely on brain injury or illness in which it is evident that otherwise healthy individuals demonstrate substantial and evidential alterations in desires, controls, appetites, sexual and spiritual orientations, beliefs and identity. Whatever that three-pound glob of matter does, how it learns and applies, interacts and regulates and so forth, our bodies are not merely a matter of circuits permitting action such as breathing or motion, but rather we are our brains in a very substantive and real way. Our "freedom," "identity," "consciousness," "sexuality," "moods," "appetites" and a long list of descriptors are all related physically and nonreducibly to our brain, its neurons, and its neurochemistry all the way down to the quantum. To paraphrase an infamous interaction "it is brain all the way down." Of course, this is also understood in the most basic and now most common medical

my belief that Jesus Christ is the incarnate Son of God has consequences for all aspects of my life." Polkinghorne, *Quantum Physics and Theology*, 13.

8 See, for example, Nancy Morvillo, *Science and Religion: Understanding the Issues* (Oxford: Wiley-Blackwell, 2010), pp. 152–56; and Michael Ruse, *The Philosophy of Human Evolution* (Cambridge: Cambridge University Press, 2012), pp. 1–38, esp. pp. 17–19. Both works, and many others, argue that the idea of evolution was a threat to the Christian perspective of creation and its idea of purpose or teleology.

description on life—as brain life—gleaned from its opposite that death is the death of brain activity however problematic that definition may be legally and philosophically. Note that I have not introduced ideas such a "mind" or "soul" but have simply mentioned phenomena that are clearly observed, via scientific method, as being related to brain activity. Persons with brain illness or injury do demonstrate alterations of personality or self, even to things as ephemeral as "belief" or social structures. Further, neuroscience has located centers or regions that explain psychopathy or aberrant social and moral behavior, areas impacted by hormonal activity early in embryonic development that predict sexuality and note that specific centers are indeed related to language acquisition and so forth. Increasingly it seems as if *I* am my brain.

The responses to this issue are varied and often flavored by the discipline from which the problem is addressed. For the philosopher, the options range from a consideration on dualism and definitions on substance and ontology/epistemology usually discussed as "consciousness," while the theologically inclined interact with ideas of monism and physicalism/panpsychism and the idea of the soul as somehow independent from the physically vulnerable three-pound glob of matter within our heads. The physicalist-inclined theologian tends to speak against the reduction of complexity, consciousness, agency or even mental organization. In summary, we are more than matter, and exceed the limits of our wetware of neurons. Others will argue that such physical considerations are not in opposition to an agnosticism toward ideas of soul or mind—but that these ideas are beyond the measurement of an MRI machine. Both are essentially monistic—postulating an intrinsic and necessary relationship between the ethereal/theological and the physical as being one and the same reality which cannot at any point exist independent of each other. The latter opts to not speak on the interaction except as describing what cannot be said—that the spiritual exists beyond the physical and our machines to observe it. The human being, it is argued, is a proper amalgamation of physical and "mind/soul" and we have not been able to observe the latter, but it is inferred as an ontological necessity. Others orient around the axis of dualism with some holding to a strident or rigid dualism of mind and body in which the Real is associated with the mind/soul or a kind of ontological functionalism in which the mind/soul animates the human person but remains distinct, even disembodied in various preexistent or intermediate states, from the body.[9] In distinction to the monistic, albeit nuanced on the boundaries, the dualists are

9 Joel B. Green, "Body or Soul? Questions at the Interface of Science and Christian Faith," in *What about the Soul: Neuroscience and Christian Anthropology*, ed. Joel B. Green (Nashville, TN: Abingdon Press, 2004), pp. 10–12.

much more interested in the differences between the boundary conditions of the physical world and the mind/soul reality and which properties belong to which. But both posit a necessary relation of mind/soul and body—the issue is that to look at only the body is to limit the conversation. Humans cannot remain *only* bodies, we cannot remain naturally natural. But presently, neuroscience suggests that we are only our bodies, and specifically our brains interpreting that body in the world. Or, as noted in the introduction, when my neuroscientist colleague would be asked about souls, identity or consciousness, he would defer to what he could measure or see as brain. Other questions, he would say are "my area" of theology or philosophy.

If this is the case, then what are the theological challenges if we concede that we are our brains interpreting our bodies in the world? Some readily spring to mind (pun intended). Start with the concept that what religion really holds true is a kind of ultimate concern as humane common life. This, of course, involves the relationship of freedom with morality, and specifically morality as a manifestation of human–divine concord and the consequent set of theological descriptions of justice both in this life but certainly more importantly in the hereafter. As humans are oriented to the divine Other, and to each other as a result, we must act freely toward that horizon in our common life. Failure to do so is, at the least, an impoverishment of fullness and, at worst, the warrant for divine judgment in the next life. The golden thread running across accounts, both monist and dualist, is that we exceed our natural selves, above our mere physical appetites and therefore unique in our freedom toward ethical horizons. But if we are limited instead by our brains in the world, then are we truly free? Continuing the survey, less apparent are issues surrounding the notion of "self" and consciousness, again not only in the finite, but certainly in the infinite to come as well as issues of providence and even the nature of the Divine (does the Divine have, in essence, a brain?). For Christian theology, as an example, the nature of the incarnation—long argued in the hypostatic natures—is once more problematized as what kind of special brain did Jesus of Nazareth have if he is indeed the God-man? Some may argue that this is indeed not at all an issue for theology, but rather for a particular heresy that has crept into theology in terms of dualism, and the form of dualism that comes from the Cartesian belief of a mind–body dichotomy in which there is supposed a little complete "you-homunculus" inside this vast physical machine. That homunculus is our mind/soul, and it belongs to another realm and in fact is the motion of life itself. The ancients called this "spirit" or "form" depending on your Hebrew or Greek orientation, but we have come to know it as "mind." Nonetheless, it is clear to me that at least the first centuries of Christianity did indeed struggle with the nature of the hypostasis in the incarnation and that while some claim to have resolved the issue, it remains very much in play to this day in terms of Christology. Jesus of Nazareth had a brain, but what this meant for his divinity is completely (and literally) a mystery of faith and a creed. But it is not this issue that I think is the most problematized, it is in terms of the idea of independence from body that is the most problematic—or, more subtly the relationship of mind/soul with body itself as a theological problem.

The ground in this question on the relationship of soul/mind with physical body aka brain in terms of neuroscience has undergone a rather seismic shift. Early discussions, as in the wider history of science and theology, argued for the plurality of interpretations, but here usually under the guise of some appeal to the mystery of emergence. Originally this conversation involved the claim that the scientific interpretations for complex human behaviors—basically anything above the primitive or biologically necessary reptilian brain stem—were themselves far too complicated to be able to say anything meaningful without substantial reductionism. In a nutshell, human behavior came from within and without nature, and this interaction was more than biochemistry and neural structure at every level so that the science concluded it was nigh impossible to say anything meaningful about causality. The product was greater than the system's pieces, and as such, emergence allowed for the possibility of a place of "mind" or "soul" or rather the idea that each of these concepts were emergent from the reducible parts of the brain. The brain was in the world, made of the world but neither science nor experience could explain the complexity of emerging identity and consciousness. This was a *qualitative* difference of disciplines, definitions, and descriptions. I name this position crypto-physicalism, a concession to the physicalist reduction that we are just brain but with the caveat that the sum of the system is greater than the reduction of the parts and at some point "consciousness" emerges and begins its own life. This, at a very simple level, is the "*ruach*" of spirit, the distinguishing of human from beast. Even the staunchest defender of a biblical anthropology, Nancey Murphy, worked from this principle as did her followers and emergence was the buzz word, following from the work of neuroscientists themselves. Physicalism was true, but not the entire position as emergence allowed for a "body plus" addition somewhere in the mere physical. There is, of course, a long history in the theological traditions on such distinguishing characteristics between the natural of humanity shared with the beasts and the *imago dei*. What is new is the appeal to complexity, as discerned by science itself (and this is a heritage of quantum mechanics being grossly misunderstood by theologians), as being the evidence of supervenience so that one is not held hostage to physical properties in determining classes or categories. But let us return to the options to physicalism.

Cognitive Studies and Theological Anthropology[10]

If we are just brain, then the way in which the brain processes and reflects information is critical to understand and the field of cognitive studies engages in this process. It has a close concord to neuroscience, and is filled with research dealing with perception, attention, memory, development of language, reasoning and so forth. It has both interest in the physical structures—from large regions to neurons—and in the ways in which information is relayed back into the extramental world through language and behavior. Both cognitive studies and neuroscience try to avoid the reduction of psychological accounts of human mental states that tend to understand the mind as being merely internal states influenced by external stimuli. In the older tradition of empirical psychology, the brain is merely wetware that stimuli write on according to a set of internally dictated parameters. This is often called the "computational or representational theory of mind." The offshoot of this consideration is what Daniel Dennett calls "the homunculus approach" wherein the scientist tries to reduce or decompose the complexity of stimuli backwards to a singular process or physical system.[11] This was, and remains, very effective in dealing with memory and perception, large-scale mappings and small-scale observations as to breaks in the system (such as the absence of a neurochemical due to trauma or age). However, and challenging to the field, while this reduction has great explanatory power, the brain is not in fact at all functionally decomposing information in a simply computational binary manner. Instead, the brain runs all kinds of heuristic models that seem to sidestep reduction of stimuli and look nothing like a computer that can only run in one manner or another. The physical brain works differently, it makes connections that computers (especially nonquantum) at present cannot and it in fact relies on both its interior macro- and microstructures and its external stimulations. The brain—how it actually "works"—is as important to understand as how it works to relay external stimuli. In short, the brain and its physical components actively construct and participate in the extramental manifestation of human language and behavior. This admittedly brings to the fore all kinds of issues for neuroscientists, who often can describe a phenomenon such as facial recognition as associated with brain area firing but cannot explain why faces and not toy trucks or previous dinners except in

---

10 For much of what follows in this section, I am indebted to the extraordinarily fine summation by Aku Visala, "Theological Anthropology and the Cognitive Sciences," in *The Ashgate Research Companion to Theological Anthropology*, ed. Joshua Farris and Charles Taliaferro (Burlington, VT: Ashgate, 2015), pp. 57–72.

11 Daniel Dennett, *Consciousness Explained* (Boston, MA: Little, Brown, 1991).

some heuristic manner by evoking an evolutionary advantage for the species that makes facial recognition a different kind of bio-mental process than what dinner I have had previously or whether I am lying about it.

At first glance, this does not seem to be terribly hostile to theological accounts—can it not be argued that in the modeling beyond the physical back into the world that the human soul manifests, and that this is both physical and nonphysical, or that complexity itself points to a point of emergence of consciousness at worst and soulfulness at best? This is the position of at least one significant Christian thinker in Nancey Murphy. But before we explore her position, it is important to return to where exactly cognitive neuroscience rubs theological positions, and in particular the ideas of Chapter 1.

Returning the Eden: Theological Anthropology

The argued foundational ideas carried in the Judeo-Christian tradition are (1) humans are uniquely created by God with a divine purpose; (2) humans are uniquely related to God on account of this creative action or Word of God; (3) humans are free to engage with this unique relationship, and in fact, some kind of afterlife or extended life is indeed dependent on this engagement; and (4) the account of Divine activity—whether Torah or Jesus Christ—is the evidence of the reality of the above points. Further consideration of these ideas included a series of secondary concepts—freedom, moral or virtue cultivation and ideas of alignment of will with divine purpose. Historically, this means that any theological anthropology will address the uniqueness of the image of God, an account of salvation and afterlife and how human will/emotions relate to this divine purpose.[12]

Historically, then theological anthropology has interpreted human uniqueness as the soul itself, but more recent theologians have used existential or depth psychological language to speak of the human unique soul as an openness to transcendence (Juergen Moltmann), or openness to human transcendent symbol making (Paul Tillich or J. Wensel van Huyssteen) or some sophisticated sociality as echoed in the Christian Trinity. God is social and therefore we are social. In all, the endpoint revolves around an anthropocentrism and the belief that humans somehow manifest divine will or truth as concern or orientation toward love or the like. However, cognitive neuroscience is increasingly demonstrating that this once thought uniquely human brain is in fact just one of many kinds that seem to do the same kind of thing—from near primates to our family dog. The question here then becomes how far does human ontological uniqueness extend?

12 Visala, "Theological Anthropology," 61.

Moving to the considerations of will/emotion and we encounter even greater quagmire of thought. Theological anthropologies are riddled deeply with considerations of metaphysics, and with ideas of freedom and rationality. Whether giving theological accounts of unnatural behavior due divine standard or decree (e.g., sexuality and family) or accounts of responsibility for sin and the justice of God, and the historically reduced role of emotion in the life of virtue, what one discovers in the neuroscientific discourse is that the brain does not work as a simple series of binary computations. Instead, the heuristics are redundant, dirty, plastic and even emotional and there is no clear account of "pure rationality" or even free choice. We are bounded by our physical structures, how they interact internally and externally and how they are bounded by our evolutionary history so that to think of a human person as supranatural is an oxymoron. All we are is our nature in nature as selected by nature. With such a dirty, random and clearly bounded naturalness, it is hard to consider how humans can exceed, despite our best metaphysics, our natures or our physical beings. If we are not free creatures as conceived by Immanuel Kant or Greek metaphysics, what we are is our nature. We are, frankly, beasts. It is easy then to consider how will/emotion and even choice becomes difficult to conceive even as a habit of piety let alone as the grounds for eternal damnation or reward. While humans may think that we are not naturally natural, neuroscience tells us we are exactly that or at best limits what we can say about those metaphysical ideas of freedom, ethics and so forth.

Finally, the consideration of afterlife is also challenged.[13] A deeper account will follow but the briefest summary of the challenge is this: if a person is their brain, and in the cognitive neuroscience understanding to speak of brain and mind/soul/self is to speak of the same thing, then what kind of afterlife ensures that a person has this brain to be this mind/soul/self? At the most basic, any afterlife must be embodied, and not only embodied but with *this* particular body which then becomes a series of absurd reductions—what time of my life? What particular body state? What particular macroformations of my brain will be used? Can I have the same atomic structure? And lastly, is this a merely facsimile like a Star Trek transporter or can it ever be the real body? While the pious may argue that this is an article of faith (as my students often do), the claims of modern scientific understanding make this claim of faith seem increasingly impossible given what we know of the material world, its laws of physics and our own understanding on the continual development

13 Omar Fakhri, "Physicalism, Bodily Resurrection and the Constitution Account," in *The Ashgate Research Companion to Theological Anthropology*, ed. Joshua Farris and Charles Taliaferro (Burlington, VT: Ashgate), pp. 103–12.

of neural networks, their decay, and their adherence to the laws of physics. While God may be thought of being able to do anything, to argue this is really to argue no-thing. In which case, we simply have fideism—belief in belief. And while fideism may provide some spiritual comfort, it makes for disengagement in the discourse of human life and at the end of discussion, the reflections we make apply to real people making real decisions in a medical world or a world of science.

Aku Visala neatly summarizes the issues faced by classical theological anthropology by the cognitive neuroscience:

> Theologians have rather large stakes in the game of consciousness and selves. Christian theological anthropology has tended to affirm a set of common-sense assumptions about the nature of human persons. These include identity over the course of our lives, moral and spiritual responsibility, and resurrection or afterlife.[14]

He also gives two broad approaches to answer the challenges above: (1) emergent or nonreductive materialism and (2) reaffirmed classical dualism. In the former, there is the concession that we are brains, all the way down but in consciousness there is something supranatural that emerges out of the increasingly complexity of the brain as it moves to become a "mind" or in theological vocabulary, a soul with all the baggage necessary of spiritual responsibility and afterlife. The account given then is the one that theologians wish to use in their interaction with neuroscience. Does every theologian/philosopher do this? No, but this is the response that is largely given by those who wish to account for a relationship between traditional monotheism and theological constructions of soul with neuroscience. It is, in short, their concerns working broadly in the monotheistic traditions.

A very brief excursion is the idea of panpsychism as an alternative point of view. The idea of panpsychism as a universal quality of mind is as slippery as the idea of soul in terms of definition and it's nearly as long in history. Panpsychism has a long history in philosophy, and an entire school of theology in process theology adopts much of the tenets of a panpsychist albeit in a spiritual description. Here the idea that a nonphysical substance (usually mind as order) is a scaled quality shared by all matter (hence, a rock has mind/soul but not in the same manner or quality as a dog) but the definitions therein are different. The entry points of panpsychism include ideas of consciousness (D. Chalmers) to mind (J. Haldane) to general spirit

---

14 Visala, "Theological Anthropology," 67.

(F. Schiller) and even informational theory (G. Bateson). The basic question is that if everything shares some "thing" in common, what is that common "thing" as manifest in the material universe? It could be order or conatus as energy as easily as mind, consciousness, informational awareness or soul. But generally, it is a philosophy that deals with mind and consciousness and spirit or soul and is human-centered or -shaped. In other words, it goes back from human experience. Hence, all atoms are "enminded" but not all things can be considered as "mind" so that one could exclude a rock (or can one?) but its constituent atoms are enminded. But this raises problems! Most salient to this work is that most consider panpsychism to be alien or oppositional to the concept of emergence of mind as nonreductive physicalism holds. To hold all things are mind/soul and some things (e.g., minded/souled humans but even here there is a continuum from sperm and egg to fertilized egg to embryo to fetus to adult) emerge to have mind/soul even at greater complexity are clearly oppositional statements. Galen Strawson neatly summarizes the problem when it comes to physicalism and panpsychism:

> I hold that the mental/experiential is physical, and I am happy to say, along with many other physicalists, that experience is "really just neurons firing," at least in the case of biological organisms like ourselves. But when I say these words I mean something completely different from what many physicalists have apparently meant by them. I certainly don't mean that all characteristics of what is going on, in the case of experience, can be described by physics and neurophysiology or any non-revolutionary extensions of them. That idea is crazy. It amounts to radical "eliminativism" with respect to experience, and it is not a form of real physicalism at all. My claim is different. It is that experiential phenomena "just are" physical, so that there is a lot more to neurons than physics and neurophysiology record (or can record).[15]

But he also adds that "brute emergence" is incoherent as nothing can come from nothing (or it is a miracle!), and therefore if one is to be a physicalist one must embrace order of mind as the panpsychic model: "So now I can say that physicalism, i.e., real physicalism, entails panexperientialism or panpsychism. All physical stuff is energy, in one form or another, and all energy."[16] I will

15 Galen Strawson "Realistic Monism: Why Physicalism Entails Panpsychism," *Journal of Consciousness Studies*, 13 (2006), p. 7.

16 Strawson, "Realistic Monism," 25.

answer the panpsychist (whose form is largely David Bohm's pilot wave) in the next chapter. But for now, it will suffice to allow it to stand as an option and one that many theologians working from the traditional Judeo-Christian tradition of monotheism and dualism see as oppositional and which some panpsychists see as equally oppositional to physicalism, Strawson aside. More simply put, panpsychism seems to assume the normativity of human experience of reality and to work backwards making it look very hostile to the notion that humans are special creations of God, which most theologians would want to hold as essential for the reasons of Chapter 1.

Emergent or Nonreductive Materialism

For emergence, the category "mind" or "soul" is not held to merely the physical (e.g., determined entirely by it) but it is possible to imagine that a physical reality can share the same category (mind or soul) but arise from different sources or realities. In short, while we are physically composed, we are spiritually constituted, and those attributes derive from a different set of principles than the physical. This is in opposition to the panpsychist who sees them as the same in origin. Humanity, and the confusion of consciousness with soul or identity, is physical plus. The neurosciences study the physical, the theologians study the plus. The strength of this argument is that it gives full credit, and boundaries, to the neurosciences (and a tip of the hat to the ever-popular appeal to the new physics and its mysterious multiverses where the God of the gaps reigns) and what they can say of human consciousness when restricted to physical parameters. It allows the functional decomposition or "computational" model of the brain to address how brains work, how we make memory, how we speak and so forth. Implicit concession, then, is to the fact that damage or biological interference of the macro/microstructures of the brain has correlative impacts on what we can observe of human behavior so that famously "brain injured Phineas Gage is not non-injured Phineas Gage." The brain is our wetware onto which experience is written, and the normative execution of said wetware allows us to be "me" or "you." But, and essentially, the now largely dismissed perspective of consciousness being a manifestation of increasing complexity in which we describe qualities rather than simple sensations (the so-called hard problem of consciousness) is maintained as the emergent position holds that as we study the increasing complexity of the interactions there arises a new set of properties of the brain that is the mind/soul itself, and of whose origin we cannot measure but nonetheless affirm to be true. So, the emergentist (as does the panpsychist) argues that "pixie dust in the synapses" that manifests to consciousness can make as much sense "as quantum coherence in the microtubules"[17] because what the mind/soul describes is not the same as simple reductive experiences. But "hot" or "cold" sensation (as manifestations of temperature experienced by skin for example) is not the same of moral agency, personal awareness or subjective states such as "happy" or "sad." For much of the 1970s to the early 2000s, philosophers/theologians and neuroscientists seemed to reside in an uneasy détente, the so-called issue of consciousness was too complex to allow simple physical reduction and consciousness studies emerged recognizing the embodied nature of cognition in which the physical, external and internal, looped back into the world and itself in a complex series of dirty heuristics rather than elegant reductions. Both sides whether emergentist or reductive physicalist concede to the complexity of consciousness as both a material and a philosophical or metaphysical entity. But slowly, especially with the aid of MRI machines and computational models and the inclusion of evolutionary explanation, the field has begun to shift more to a more classical reduction to the physical. And the theologians have followed suit, but as I have argued above, they have always been committed to this but with the addition of "pixie dust" to make qualities or subjective statements real.

In 2013, Nancey Murphy argued openly for a physicalist position, something she thought nearly impossible in the 1990s:[18]

> I speak on physicalism at colleges and universities all over the country [...] not so much to argue against the dualists, but to make it clear that nonreductive physicalism is an acceptable position for Christians to hold.[19]

There are several caveats to explore in Murphy's understanding—her qualifiers "non-reductive," and later exactly which kind of Christian can hold such a

---

17 Paraphrasing Patricia Churchland, *Brain-wise: Studies in Neurophilosophy* (Cambridge, MA: MIT Press, 2002), p. 197. As an eliminative materialist philosopher, Churchland argues contra this position that consciousness is a series of qualitative statements that cannot be reduced to simple or reductive biological states within the mind. While this is "common sensical" (or folk psychology) she argues that these are false illusions used by the mind to "shortcut" manifold experience and that inability to explain the propositions and subjective realities is not the same as saying they are not reducible to physical states. All it is saying is that it is not possible, and the history of theory replacement demonstrates both the logical fallacies of such ideas, and the eventual overcoming of them by better physical scientific understanding.

18 Nancey Murphy, "Do Humans Have Souls? Perspectives from Philosophy, Science and Religion," *Interpretation* 67, 1, (2013), p. 31.

19 Murphy, "Do Humans," 32.

position as she notes later in her article that neither "Catholic nor Calvinist" can indeed find concord in her theological synthesis.[20] But, presently, it is enough to note that Murphy feels, as she goes to note, that physicalism is the only rational position to take given the state of neuroscientific research (and in agreement with Strawson). In particular, she notes: (1) the increased understanding on the relationship of brain to psychopathy and behavior; (2) studies on the relationship of brain to freedom and motor choices (particularly in the dysmorphing of choice and identity with motor function); and (3) finally studies on brain injury and personality/belief in which there is demonstrated alteration of self-awareness or assertion by individuals having undergone brain injury or trauma.[21] One, she notes, could also add research on the relationship of language (as manifest in specific cortical locations and development) to symbolic self-concept and a host of studies on the notion of embodiment as a function of brain activity or absence.[22] She also notes that our entire Western sense of legal and moral responsibility rests on the notion of freedom and identity, that this may be a location of much controversy but the evidence seems to suggest that self-concept is dependent on physical structures being adequate (whatever that may mean), and that this is not necessarily a challenge for moral or legal responsibility.[23] Murphy cannot help but conclude that the simple axiom "I am my brain" seems more and more certain as a basic truth because at every level it is impossible to separate the "I" from "brain"—they are one and the same. Add to this comparative neuroscience, and the lack/presence in the animal world of certain brain structures, even in near primates, as explanations for behavior and we see further evidence of the relationship between the brain and the self, soul or mind (let alone a blow against those who hold for a pan-soul across all life or some kind of animal soul).

In addition to this startling admission, Murphy also extends the conversation in another direction which is to argue that the dualist position long associated by Christian theologians, particularly in the West, as being indeed the normative position is, itself, a product of an aberrant thought process not current with either Hebraic or Christian Scripture or a host of doctrinal understandings, including the nature of the afterlife.[24] She, to my knowledge, remains agnostic on what exactly those doctrinal shifts may mean but she signals something

20 Murphy, "Do Humans," 34.
21 Murphy, "Do Humans," 34–39.
22 Murphy, "Do Humans," 40.
23 Murphy, "Do Humans," 41.
24 Murphy, "Do Humans," 41.

different—simultaneously a critique of mind/soul and body dualism, a re-thinking of biblical interpretations toward some kind of pantheism and an admission that the emergence theory is not quite dead but finds new life in consilience. Each of these will be dealt with in turn.

A Critique of Dualism

Human nature or human identity? Are they the same thing? Have we privileged one or the other? Certainly our contemporary philosophical situation, as in Sartre, has rejected the idea of a monolithic human nature in favor of a personalized individual authentic self and we have moved to discuss interiority, again starting with Descartes but even with Augustine's understanding of memory, as primal self-understanding.[25] Carried in the enkratic traditions on the pastoral care of the soul after the gift of baptism or as part of spiritual maturity, it is normal to speak of one's soul as separate from the body in some mode or another, as part and parcel of soul care is the denigration or subjugation of the body. Thus, in both Christianity and Judaism there are traditions of self-denial of appetites of the body to enhance soul care. Further, some philosophers such as Richard Swinburne argue that to speak of human bodies without souls is an impoverishment of things that we value most of all—such as sacrifice, altruistic love—and particularly the denial of the body for such virtues. Thus, a heroic first responder who sacrifices for another or mother who starves her body for her child is proof of the priority of the soul as a value—it remains after the body is lost and is the source of our willingness to lose/harm our bodies.

Of course, this is entirely mirrored in the Jewish and Christian traditions. In both, there are concepts of afterlife and resurrection. For Jews, dualism is less a problem as the claim is that the body itself is generally resurrected but the soul has no intermediate state between life and resurrection. Bodies are souls; there is no soul without a body. But for Christians, it is much more problematic when it comes to the incarnation when one considers the technically intense discussions between anhypostasia and enhypostasia when thinking of Jesus of Nazareth and the resurrection (both his on Easter Sunday and ours at the end of time). Does he possess a general human nature or a unique human nature of which the Godhead annexes or lifts to self? It may seem like splitting hairs, but the orthopraxy implication is real as anhypostasia means that there is a

25 Nancey Murphy, "Introduction," in *Human Identity at the Intersection of Science, Technology and Religion*, ed. Nancey Murphy and Christopher Knight, (Burlington, VT: Ashgate, 2010), pp. 2–3.

general humanity lifted into the Godhead but that it really does not matter whether it is a Darren or a Sally. Identity is irrelevant. It leads to thinking tending toward Docetism or general Platonism. Jesus only looks like a human but is only a human wearing divinity or one needing to complete itself in the divine. Enhypostasia allows for this personal identity but tends to lead to a view that is completionist or treats humans as ontologically incomplete (e.g., sinful) before joined with God (presumably in baptism). Here, the stress is that until one is "spirited," one is somehow not quite human—a Frankenstein monster waiting spiritual-ification. For many Christian theologians, this happens at baptism and the pre-baptized state is a kind of monstrous variation which, I may add, explains the necessity of both immediate infant baptism and the various decisions such as who can be buried where. Once the Protestant idea of election is added, then you are either thrust into a universalist position or a hardcore Augustine/Calvin election position in which the world is in essence two humanities—one reprobate and one elect. The elect are the only real humans—selected with spiritual lives—while the reprobate are merely monstrous facsimiles, easily disposed by society (Puritanism) and God.

One can, of course, press fast-forward to Descartes and his insistence on an individual independent of the body in terms of mind and we have the same heresy as Origen proposed and condemned at the Second Council of Constantinople in 553 CE.[26] Fergus Kerr notes that this is merely a simple Wittgensteinian linguistic transposition in which "I" am my linguistic construction, my inner self independent of body or matter.[27] To speak of "I" as some kind of preexistent spiritual soul before bodily consummation is a heresy. Likewise, resurrection is problematic as most Christian theologies hold to an intermediate state between physical death and resurrection. The soul, it is claimed, exists without a body for some period of existence although properly a person is soul and body as in the resurrection. So, in the incarnation, the resurrection of Jesus of Nazareth and the general resurrection of the faithful dualism is affirmed but not without problems. Whether classically or in near modern terms, dualism is still problematic. But it is also true that both Jewish and Christian traditions are largely dualistic, despite historical doctrinal objections such as the above.

But why does this matter for neuroscience and physicalism? Can it not be thought that the collapse of the self into the physical is in fact not only welcomed to solve the old chestnut of dualism? Can it not be thought that to consider ourselves as both mind/soul and body simultaneously is better than to try to split the hairs between them? Certainly, the moral theorists such

---

26 See Fergus Kerr, *Theology after Wittgenstein* (London: SPCK, 1997).

27 Kerr, *Theology*, 23.

as Fergus Kerr think so—Kerr suggests that the problem with moral theology in the last few centuries—whether Catholic but especially Protestant—is that deciding between act and intention is now a matter of linguistic analysis ("I" versus "me" in which "I" is the pure self and "me" is the world-infused fallen self) rather than the manifestation of action that is the bedrock of Catholic moral theology for most of its history and certainly in its real sacramental application in terms of venal and mortal sins.[28] Is sin in thought, word or deed the same thing? Of course, note that we have not even breached the idea of freedom, even circumscribed freedom in the life of grace following baptism for original sin. Kerr's rejection of the linguistic structure of identity because it reveals a tacit belief in an ontological description of soul and soul plus body is telling. In the last chapter, I will argue that it provides the best ground for a rapprochement of theology to neuroscience using Freud's ideas of positive fiction over negative illusion. We use theology to look at ourselves, it was once the best fiction for us to use but now is an illusion. A new fiction is needed. This is being naturally natural—a nature realist and the best mind map is that of science.

Dualism is a problem. The solution, at least for Murphy and her progeny, presently is physicalism married with emergence or consilience appealing to quantum theory as allowing "homo religiosus" to exist within all the probabilities of religious expression and plurality. As Christopher Knight argues the solution to neuroscience's restrictive physicalism is a mode of his own "strong theistic naturalism" (which he admits can be easily read as Deism) into which five critical points flow. Here is the pantheistic anthropology: (1) human psyche/soul/mind is a natural biological process from the Big Bang through human evolution; (2) there is no need to enact any special divine action to explain human experiences including divine presence; (3) it is almost universal that such human experiences of the divine are common; and therefore (4) a diversification of said soteriological and not propositional/dogmatic (borrowing from Keith Ward) manifestations of divine experience in disparate historical manifestation is expected; and (5) the fruit of a pan-cosmic will to make these responses is the foundation or real.[29] This is, of course, a theological rehash of Michael Polanyi and Roy Bhaskar's critical realism in which it is argued that perception is foundational to discovery of objective foundational reality. In short form, we are hardwired to perceive ontology, use language to discuss it and this perception and codification does not in any way

---

28 Kerr, *Theology*, 127.

29 Christopher Knight, "Homo Religiosus: A Theological Proposal for a Scientific and Pluralistic Age," in *Human Identity at the Intersection of Science, Technology and Religion*, ed. Nancey Murphy and Christopher Knight (Burlington, VT: Ashgate, 2010), p. 30.

alter the reality of the Real. For theological/religious statements, as in scientific statements (in terms of equations), they are genuinely referential to true structure despite any infelicity in articulation or expression.[30] Arthur Peacocke makes this a feature of his largely psychological-referential model of religious language, while Simon Conway Morris does it in terms of human evolution but both—or all including Knight—end with Polkinghorne's world infused with mind.[31] What none really bothers to mention, also in Alister McGrath, is that this is borne out of Bhaskar's reduction of physics as understood in the quantum as being such a form of critical realism.

Why does this matter? Well, it matters because if the metaphysical "wave package" is increasing and collapsing to the physical—as predicted and admitted by Murphy now—then we cannot possibly hold to a "strong theistic naturalism" as the psychological referential system as the location of mind/soul is now quantum-level mechanics because the natural is completely natural. The point of the pan-entheistic mind is now the quantum mystery itself—the last undiscovered country. To the point of the quantum interactions within neurons, we are merely machines working. Is this a problem?

A "New" Emergence Theory

As mentioned, even the staunchest defenders such as Nancey Murphy of nonreductive physicalism have recanted their positions, but she has not yet yielded the field completely as she still holds to an emergence theory based on consilience in positing that while it may be true that we are just our brain when the complexity of the brain is considered with respect to consciousness and identity, the sum of the parts is greater than the whole. This is not an appeal to distinction between "mind" and "brain" at first glance[32] but rather a more complex appeal to "dynamic systems [that] are often the source of their own changes" of which "some behave deterministically, and others are affected by genuine quantum-level indeterminacy" so the end point is that "some systems are causes in their own right."[33] What Murphy posits is that while the system is completely physical and therefore determined, because of *quantum-level interactions* it is possible to think that these systems can exceed the limitations of determinism and in effect be self-causing or free. Even more

30 Knight, "Homo Religiosus," 33.

31 Knight, "Homo Religiosus," 34.

32 See Nancey Murphy's summation in Nancey Murphy, "Reductionism and Emergence: A Critical Perspective," in *Human Identity at the Intersection of Science, Technology and Religion*, ed. Nancey Murphy and Christopher Knight (Burlington, VT: Ashgate, 2010), pp. 79–96.

33 Murphy, "Reductionism," 96.

simple, consciousness is more than its parts, and is its own self cause as it arises—emerges—out of the mystery of the quantum world. The physical world does not matter then except to fix the possibilities of the quantum as a fly is caught in amber. The nature of the fly is itself the field of study, the amber is merely a fixing in time and space. Thus, in this analogy, the brain is the amber that captures or fixes the fly, which is consciousness or the soul. The fly is the object of interest, not the amber. The fly is the mystery of the quantum world.

Now, let us think carefully on this. There are several steps in this argument for emergence. First is a critique, as old as the epicureans, on atomic reductionism in which it was thought easy to parse out human behavior, and all reality, into atomistic deterministic components so that any complex reality or action is a sum of its parts. While this is a desirable idea, and perhaps has indeed yielded much in terms of scientific discovery and application, when one looks at many things, including human behavior, it becomes very difficult to maintain. This is "upward causation"—things start most basic and complexity is all the way up—and emergence steals from evolutionary and natural biology's "downward causation" in which environment alters fundamental physical structures (and therefore causation) so to alter substantially the behavior and reality of the animal as a result. Hence, a food alteration impacts a sexual characteristic that alters the selection of some families selecting for certain traits, which leads to speciation. A critique of this misappropriation is beyond this scope, although suffice it to say that it is not downward causation as the characteristic is not *ex nihilo*. For emergence theorists, causation and complexity, unlike biological downward causation, comes from deep structures preexistent in the soul of the person—literally out of nothing (recall Strawson's objections). For downward causation in evolutionary biology, environment acts on physical structures in a dialectical fashion which then, in turn, influences what the animal experiences anew and it is, indeed, impossible to predict from first principles how the complex organization of those interactions will occur. For the emergence theologian, biology as well as the quantum, tells us that there is more to heaven and earth than dreamt in our natural philosophy of simple reduction. But the point is that they are not the same things at all.

Emergence also steals from systems theory, and the work of the philosopher Alicia Juarrero, by arguing that proof from nonequilibrium thermodynamics (and one would be wise to recall that it was thermodynamic problems that led Ludwig Boltzmann to reinterpret James Maxwell's work on thermal equilibrium, which in turn led Planck, Einstein and Bohr to birth quantum mechanics) as understood by pattern theory predict that complex systems exceed and even self-cause beyond its component parts. Using common sense—say identifying that a roll of the die is context free and its probability unimpacted by successive rolls, whereas as a deck of cards with one being drawn has context influenced probability depending on which cards were drawn—Juarrero concludes that the reality of what has happened flows downward to impact reality. The higher level—this card drawn—impacts the lower-level reality by changing the probability of a single card being drawn. Juarrero also steals from thermodynamics and the reality of the fact that most natural systems (such as the atmosphere) of thermodynamics are in a state of flux beyond the scope of classical thermodynamics to explain as the system is too complex, let alone the issues addressed by classical physicists and solved by Planck et al. What the emergence theologians want to argue is that the phenomena of consciousness—higher-level realities—flow backwards to explain lack in the lower level of simple neurological processes and brain functions but when the scientist studies only the neuron—all they can observe is the lower level itself. Causality has flown backwards as consciousness determines its own reality in its entire complexity. This is what Juarrero argues, and what Murphy intimates. But is the complexity of the system, the failure of linear classical thermodynamics to account for all the realities in the system and to predict it, the same as arguing "they become their own causes"?[34]

Certainly, those who have won Nobel Prizes in nonequilibrium thermodynamics such as Lars Onsagar and Ilya Prigogine would object as it is the point of their work to integrate classical and nonequilibrium thermodynamics by not dismissing completely classical mechanics but rather by extending it using time as nonlinear. The issue for Prigogine is the intrusion of linear time, refuted at least three times from Newton's Euclidean space, and instead launches an appeal to systems theory that deny this basic structure as we perceive it.[35] In essence, what Prigogine argues is not that the top influences the bottom, or the bottom influences the top, but that the reality *perceived* is a problem because at the quantum level the idea of causality as successive moments of time is nonlinear. The quantum is different but not mysterious or self-causing; it follows its own lawfulness, which is different from the manner through which humans perceive reality, especially in time. But this is not the same as saying that there are not fundamental laws that determine this causality or that this is the realm of God, rather our tools are not presently mathematically or theoretically sufficient. If it were otherwise, then rather try to understand, we should just sit back and enjoy the show as on occasion elephants turn into unicorns and broken glass reforms into bottles.

34 Murphy, "Reductionism," 91.
35 Ilya Prigogine, *The End of Certainty* (New York: Free Press, 1997).

It is this last point, the odd confusion or appeal to mystery that drives the argument against physicalism and instead pushes for a kind of emergence/consilience argument. It is usually made by the appeal to the quantum, expressively so in Murphy and in Juarrero, where the mystery and the inability for classical models of physics are used to justify what is, in essence, a leap of faith. It is, for my money, a playground argument that my Dad can beat your Dad and one which theology and religion has a long history of hiding behind and in which only, in that history, to find itself outstripped by its scientific sibling. Whether an appeal to the Kalam cosmological argument that something cannot come out of nothing in terms of the Big Bang or that evolution allows for an ensouled animal called "humanity" to arise, the point is that sometimes theology in its attempt to correlate itself sees in scientific discourse the hand of God hidden usually as mystery. And then sometimes, the arrival of many different hominids on the evolutionary scene or the rise of m-theory in cosmology suddenly makes the mysterious less mysterious and more under the traditional banner of classical mechanics.

What one has, in my opinion, is a third-order emergence theory in which it is claimed that traditional bottom to top organization and influence not only fails to explain all phenomena (first-order emergence) but also that the limits/laws of the bottom organization are exceeded by the system so as to be said to be self-organizing (second-order emergence) but more importantly that there is some unknown principle—teleology—at the top levels of system that flows back to the lower fundamental systems.[36] This "reality" or "mystery" is the new realm of the Divine. It is, as it is explicitly or implicitly claimed, despite the fact that those who work in this field would have an allergic reaction to this claim, the realm of those theoretical physicists who specialize in quantum theory and in particular, m-theory or string theory or by evolutionary biologists who would argue that Terence Deacon's claim of teleology of heart muscle is greater than the atomic organization of such proteins and so on is patently a misappropriation of the idea of evolution itself and just a rehash of intelligent design. So, the first thing we need to consider is whether the quantum is indeed a mystery that has teleology and a queer teleology as organizing principle that exceeds the sum of its parts?

36 Murphy, "Reductionism," 93–95.

Chapter 3

MULTIVERSES OR THE HOME OF GOD?

The claim that there is a different universe, or lack of fundamental laws of physics, in the quantum or in the cosmological is false, popular and a misunderstanding of what most physicists and science understands the quantum as being. It is different place—outside of *our* perception of time and space—but it is not a mystery or random, as Einstein famously worried (or Feynmann joked). Mystery, rather, in terms of quantum mechanics needs to be understood under the rubric of Heisenberg's Uncertainty Principle and remains scientifically agnostic. What theory predicts, and evidence suggests, is something different, and certainly hostile to the traditional understanding of God in the Abrahamic traditions. As Sean Carroll writes, the problem is not what we do not know in terms of the quantum but rather that we know enough justifiably to argue that the core theory itself seems unassailable. It is just not that it is "misused and abused by cranks and charlatans—and misunderstood by people struggling in good faith with difficult ideas"[1] but that it is audaciously simple and powerful in its scientific and real value:

(1) Everything we know says that quantum field theory is the correct framework for describing the physics underlying everyday life.
(2) The rules of quantum field theory imply that there cannot be any new particles, forces or interactions [including God/souls] that could be relevant to our everyday lives. We have found them all.[2]

It is the second point that is the most important: we have found them all at least in the local universe we live in. Is this a statement of incredible pride or is it true?

What then does this all mean?

---

1 Sean Carroll, *The Big Picture: On the Origins of Life, Meaning and the Universe Itself* (New York: Dutton, 2015), p. 160.

2 Carroll, *Big Picture*, 179.

It is clear by any account of quantum universe cosmology that the universe is much more complex than predicted by the standard account of physics and that it is likely that the laws of physics are varied not only in terms of other universes but perhaps even within "our" universe (assuming that we are a bubble in a much larger universe as some predict).[3] While classical physics required constants—20 odd in the standard model—that are normative within any given region of space, the possibility is raised in quantum chromodynamics cosmology (or other core theories such as quantum field theory) is that these constants may be varied, depending on the presetting conditions of the creation of that universe. For example, variation in the Higgs field would dictate variation in the mass of fundamental particles. But does this mean anything goes? The collapse of this multiplicity of possibilities into a single set of fundamental constants that determine the causal relationship with the physical universe can be explained by a core theory of quantum mechanics as developed over the last century and as expanded into cosmology and other fields. What began as a study of the problem of measurement (aka the measurement problem) in the very small and observation of light or other radiations has expanded to almost every scientific field from cosmology to biology. This is also important for understanding how *emergence* within the quantum cannot work as claimed in terms of neuroscience wherein it is thought the complexity of processes exceeds the limitations of the system because of the quantum nature of all matter interaction. Consciousness is merely quantum mechanics in the nanotubules of our neurons. Whether emergentist or panpsychist, is not the mysterious world of field theory and pilot waves the truth of God and our souls?

What Is Core Theory?

Before exploring the development of classical quantum mechanics, as this history is a source of popular misconception and the place of the "God of the gaps" with terms such as "spooky action" or entanglement and ideas such as a dead cat, we should explore the terrain of present quantum theory. As much in theoretical physics, this is a very obtuse area. The mathematics involved are bizarrely abstract, and the subject matter of study, by definition, is clearly other than our physical universe in terms of behavior and our ability to measure or gauge it. The physicist "sees" it numerically or mathematically and measures the quantum world obliquely as it brushes against our universe. Furthering the complication is that quantum mechanics must at some level agree with special

3 Paul Davis, *The Goldilocks Enigma* (New York: Mariner Books, 2008), p. 153.

relativity.[4] As Adam Becker's recent work argued, there is much unfinished work, and more importantly, unfinished metaphysics in the meaning of quantum physics.[5] The merger of these technical problems and two theories in quantum mechanics and general relativity is core theory. Most physicists agree that there is, or should be, a core theory—quantum field theory which brings together general relativity and quantum theory—of which there are two present candidates in conventional quantum field theory (CQFT) and axiomatic quantum field theory (AQFT). The reason for the two candidates involves decisions of both mathematics (especially on the nature of Hilbert space), how to "renormalize" the particles we come to know in the other dimensional field in our physical world and finally questions on prejudice in selection of tools to measure the predictions.[6] However, CQFT has increasingly proven much more powerful than the alternative proposal. But the basic criticisms remain, that our decisions before even mathematical exploration can lead to contradictions within the framework and impossible Eigenstates[7] in the mathematics.[8] For some, but not most, physicists this is counter to the scientific method and its claim of simplicity, continuity and experimentation (and presumably reversibility (e.g., we can go back to other states of any system). For others, as exhaustively argued by Adam Becker, it is due to an overcommitment of physicists to logical positivism[9] and, by implication, a

4 For an excellent summary on the necessary coexistence of quantum mechanics and special relativity, see Michael Redhead, *Incompleteness, Non-Locality and Realism* (Oxford: Oxford University Press, 1988).

5 See Adam Becker, *What Is Real? The Unfinished Quest for the Meaning of Quantum Physics* (New York: Basic Books, 2018).

6 See, for a brief summary, Christopher Smeenk and W. C. Myrwold, "Introduction to Philosophy of Quantum Field Theory," *Studies in History and Philosophy of Modern Physics* 42 (2011), pp. 77–80.

7 Jeffery Bub has an elegant work dealing with the measurement problem as the Eigenstate–Eigenvalue link in which it is assumed that the statistical dynamic values of a quantum mechanical system can be detected via classical modes. This is the heart of the "measurement" problem identified in the Einstein, Podolsky and Rosen (EPR) paradox and thought mostly deeply by John Bell. The issue is going backwards—from determinate values (e.g., real numbers or observable properties of spin) to the indeterminate quantum independent of both time and observation. This makes the fine-for-all-practical-purposes (FAPP) proposal particularly as decoherence both suggestive and powerful while also limiting it. Jeffery Bub, *Interpreting the Quantum World* (Cambridge: Cambridge University Press, 1997).

8 Laura Ruetsche, "Philosophical Aspects of Quantum Field Theory," *Philosophy Compass* 7, 8 (2012), pp. 559–70.

9 Becker repeatedly makes this explicit argument on the deficit of the Copenhagen Interpretation running from Einstein through to Anton Zeilinger. Becker, *What Is Real?*, 267–85.

deficit of imagination to alternatives such as Bohm's pilot wave and Everett's multiverses. Some of the proposed Eigenstates of particles in CQFT are none of the above.[10] Olivier Darrigol neatly summarizes the limitations of the mathematics and theory/ies:

> We should content ourselves with the insight that quantum discontinuity [in terms of abstraction to real world axioms], if it is admitted as a fundamental feature of the microworld and if it is complemented with natural axioms [decisions] concerning the relation between micro and macro world, necessarily leads to quantum mechanics [as CQFT and its various candidates] as we know it. It remains possible [as in the pioneers of quantum mechanics who worked against the prejudices of the classical physics world] that someday quantum mechanics will be regarded as an approximation or a special case of a more fundamental theory.[11]

Despite the various concerns, what we have works—and explains more than it does not. It is, as discussed in a later chapter, not necessarily an ontological foundational statement but rather a semantic theoretical statement that is characteristic of science over the abrupt fundamentalist ontological statements of scientism. It cannot skirt toward metaphysics—the very thing the so-called logical positivism felt carried in the bones of the Copenhagen Interpretation and its heirs in CQFT wishes to resist. But it is also that this is a wonderful place for theologians to revisit the popular claim that at the end of physics, and quantum and special relativity integration they arrive to find God sitting there waiting for them.

It must be admitted, and risking adding to the mystery that ineluctably draws philosophers and theologians to a "God of the gaps" seduction, that neither quantum mechanics or quantum field theory have one singular massively agreed shape to deal with all the issues of nonlocality/entanglement and measurement issues or that special relativity is singular in its Einstein classical formation. Nonetheless, all the candidates do a similar thing—posit mathematical interpretations that predict real-world events, and then test for them. What is common is the sheer validity of the shape of the theories. What remains, and perhaps will always be such, is that there is much more applicable than not, and certainly in terms of the prequantum revolution.

---

10 On the reasonable axioms of quantum mechanics and CQFT, see Olivier Darrigol, "Shut Up and Contemplate!: Lucian Hardy's Reasonable Axioms for Quantum Theory," *Studies in the History and Philosophy of Modern Physics* (2015), pp. 326–42.

11 Darrigol, "Shut Up," 341.

Sean Carroll's recent book, *The Big Picture*, is an excellent source to start with as he neatly (and mathematically) summarized his "core theory" of quantum mechanics and relativity. This is an amalgam of "quantum field theory," which accounts for the most basic issue in quantum mechanics—that of observation or measurement and probability—and its relationship with our very stable world of constants in classical physics, chemistry and even biology. In its most simple form, any quantum field theory argues that all reality, and specifically the reality of the quantum, is described (as below) as a wave function and that this is essentially a derivation of the Schrödinger equation. It is the evolution of any quantum field into which all things, including observers of a quantum event, that matters and there is no special role for observer or system except that of the field itself. The measurement problem of the Copenhagen Interpretation, as Bohm, Bell and Everett suspected, is entangled in the wave function itself so that there is no boundary between the quantum and our world which one magically crosses. This means not only does quantum mechanics describe the event of the collapse of superposition, but it also describes the observer in the field itself. We are entangled, placed into the wave function itself as it evolves, so that there are no new variables or possibilities but those we are observing at this moment, this time, and this event. This is perfectly "classical" in that it is deterministic in that if we have enough measurements, we can predict what will happen, but the problem is that we are in fact in the experiment rather than observers. This is sometimes called the "many-worlds theory" because each event of the wave function in the field is itself a "universe." There is a logical metaphysics in this—each collapse is therefore a new branch of reality, and this is, by definition, infinite. But this, as we will argue below, is a problem of language rather than concept. What is important are two absolutely critical concepts: (1) core field theory is compatible with the physical world, is real and is deterministic so that the future follows from the present, and (2) what we see when we are looking at the world (the wave function collapsed) is the same as the quantum when we are not looking at it (the wave function in superposition).[12] This answers Einstein's great worry that there is a logical inconsistency in the Copenhagen Interpretation and the measurement problem—namely an artificial line between the real of experience and the real of quantum mechanics, or gravity for that matter.[13]

12 Carroll, *Big Picture*, 170–71.

13 Becker, *What Is Real?*, 13–88, 193–218.

What this means is that our world—of particles behaving like waves as four fundamental forces—arise out of fields that stretch out across all points in spacetime and that this basic field(s) make up reality including us. These fields, in turn, are described not by position in space but as possibilities (superposition) in a wave function that its member particles can assume. Particles are broken down into two fields—bosons and fermions—of which all the known forces in the physical world manifest from. The boson field has unique properties and can combine to create a unique kind of field or wave function that is associated with gravity and electromagnetism. The fermion field is also unique in that these particles (along with the Higgs field) occupy spacetime and follow the Pauli exclusion principle so that only one can be in one place at any time, and these make up electrons, protons and neutrons and are associated with the weak and strong nuclear forces but also account for matter itself. Both fields are combined in core theory (or in string theory) and are combined with general relativity when one pushes outside our normally slow, cold, and large universe (e.g., early cosmology or black holes). Core field theory "works" in our universe with its particular manifestations of the fermion and boson fields. In fact, it is the astounding predictability and reliability of this theory that gives it strength—no new particles have been accounted for, and all those previously predicted are being found as in dark matter. There has been no evidence of new fields in any crossing symmetry now for some 40 years of observation in our universe. There may be new particles—and this is the story of black holes—but these will be not relevant to our world except as mathematical predictions. Those particles, such as dark matter, that we expect to find are, ultimately, irrelevant to our physical universe as they interact so weakly with the two primary fields that they do not have any impact on our world. They are "there" but not necessary at this juncture except as predicted by the theory itself.

Sean Carroll, again, is helpful to dispel any mythologies that might surface with such imperial statements on core theory or its limits. It is really an effective theory—good at looking at the wave fields, including us in them, at the limits of our long-wavelength (distant), low-energy (cold) universe:

> In quantum field theory, the modifier "effective" does not mean something like "does a good job fitting the data." Rather an effective theory is an emergent approximation to a deeper theory. A kind of approximation that is specific, reliable and well controlled—all due to the power of quantum field theory […] the actual ontology of the theory can be completely different than the effective theory. That's the multiple realizability in the context of quantum physics […] but there

> are not that many different ways that quantum fields can behave at low energies.[14]

What we have then in the quantum world *is* our world. It is our world as we have examined it, understood it, tested it and predicted scientifically with the most rigorous methods we can devise, and it has proven effective. It explains "us," life and the universe. The challenge for faith communities is not whether quantum mechanics is true or not, but rather to them that the effective nature of quantum field theory is the most adequate, powerful, simple way to explain our world including our minds, souls and spiritual lives:

> Quantum fields [as we are made up of stuff], however, are indisputably part of who we are. If they are *all* of who we are, we should have no problems drawing out the implications of that fact for our lives. If there is something in addition to the quantum fields, it is reasonable to seek an understanding of (and evidence for) that something that is just as precise and rigorous and reproducible as the one we have for field theory.[15]

Isn't this then exactly what the panpsychist and emergent theologian both argue—one continuum of reality in which our final soul destination is the most complex of that fundamental reality? Is there a ghost in the machine of quantum mechanics?

A Basic Quantum Mechanics

Quantum mechanics and the "new physics" (sometimes called quantum physics (QP) or quantum theory (QT)) is often the refuge for theologians and philosophers to place mystery and to find concord in both method and content between theology and science. Whether as a place to play with paradox or survey of the incomplete heuristics of methods, the new physics has long been a resource for theologians to source theological convictions such as faith, accommodation to human understanding of truth and the like.[16] Citing critical realism, advocates argue that the new physics (and in particular, the Heisenberg uncertainty principle (HUP)) was itself an exercise of faith in that the physicists conjured constants out of the air to solve mathematical issues such as done by Planck, De Broglie and others, which only later were

14 Carroll, *Big Picture*, 186–88.
15 Carroll, *Big Picture*, 220.
16 See, e.g., John Polkinghorne, *Quantum Physics and Theology: An Unexpected Kinship* (New Haven, CT: Yale University Press, 2007).

confirmed by experiment. They dreamt the quantum, and then designed experiments to prove it to be true is the basest charge. Further, it was a matter of accommodation to human understanding as the very small (or very fast) was inconceivable to human minds in models—whether electron orbits as matter waves or wave–particle duality in light—and the temporality/causality of the quantum world was different than experienced by humans. Our minds accommodated reality but are not the Real (including the supernatural) but are representations of that fitted for our understanding, even if the mathematics was sound. How could any mind really understand the matrix mathematics of Max Born applied to real atoms? We can observe the orbit of Mars around the Sun, but can we understand the quantum interactions of unimaginable numbers of atoms, each constructed of electrons and neutrons, each constructed of quarks and bosons, spinning and vibrating individually to produce fields and matter that make up such an easily mathematically simple event such as a planetary orbit in spacetime? It can be symboled through equations of Schrödinger and played with by Feynmann's famous comment that the entire universe mathematically could be but one electron at all places in time at once to demonstrate the probability of the quantum world. Finally, as John Polkinghorne argues, it is also ultimately an act of humility—placing aside one's own subjectivity and bias to see objectively what the universe is revealing.[17] The universe, according to him, is "shot through with signs of [Divine] mind" so that theology and science both must put aside any hubris of interpretation to read the signs,[18] presumably assuming this mind wishes to be revealed. Both are presumably public—one in the academy and the other in tradition—and both suffer from periods of certainty and confusion that lead to synthesis and development. But is this a fair reading of the quantum?

As Adam Becker's work argues, in part, this is a historical problem associated with the heady days of the early twentieth century, European politics including the rise of communism and fascism and the rise of the United States in terms of physics following the Manhattan Project.[19] Logical positivism, and its near cousin in materialism made concrete in Marxism, rebelled against the role of Church and aristocratic state in which God ordered the universe along seemingly arbitrary divine will. Logical positivism and materialism in Marxism promised a bold new world of reasoned order, of decisions (personal and political) not based on a divine fiat but rather on discursive argument. Thus, physics found itself also embroiled in the same commitments as the sociopolitical and economic worlds. The Copenhagen Interpretation then

17 Polkinghore, *Quantum Physics*, 6–7.
18 Polkinghore, *Quantum Physics*, 9.
19 Becker, *What Is Real?*, 165–89.

is just a positivist materialistic legerdemain that infected all of the Western rational metaphysics. It resisted any description of reality, including quantum reality, that did not easily fit such reduction, overlooked mathematical problems and ignored critics such as Einstein. What Becker calls scientific realism (Quine) is a more modest claim—namely, that science gets much correct, but it is indeed a human activity and it is prone to bias. However, in later chapters it will be asked whether this is enough to allow theology to enter the discussion of the Real? For now, it is enough to agree with Becker and the dissenters to Copenhagen that the early quantum scientists were indeed infected with human failings, history and commitments. But does this mean that Polkinghorne is correct in allowing God to be the pilot wave? A rather lengthy exposition on the development of quantum theory is necessary as it is often told as a solution to the impossible or impossibly observed and as such, a kind of alchemy in perception that was broken by complete intuition into deeper mysteries.

Early in the heady days of the development of the new theory Max Born found a way to describe Schrödinger's and Heisenberg's work as it related to well-known measured phenomena such as the photoelectric effect. Born's solution treated Schrödinger's Ψ wave function as a concept of probability that has no one true physical reality, unlike the electromagnetic field observed and measured in the photoelectric effect. Schrödinger's famous cat thought experiment (his own objections to Born's solution) became the gold standard explanation (later melded with the HUP) for Born's hypothesis—the superposition of two or more states (dead or alive cats as a probability wave) is collapsed by the act of observation itself. Superposition is a probability wave function, but observation fixes, via the HUP, the wave effectively collapsing it to one location and a series of physical properties in spacetime. This is commonly referred to as decoherence. Out of a nearly unimaginable wave function of possibilities (Eigenvalues), a singular Eigenstate emerges and this at some level is induced by the environment of measurement/observation.[20] Certainly some (e.g., Eugene Wigner who thought John von Neumann's theorem demanded this) gravitated toward an understanding that consciousness itself collapses the wave function. But this raises the question, who collapses the wave function—a human mind, a dog, a flea?[21] The folding of light (electromagnetism) into the consideration by Dirac's quantum field theory adds another important concept as the tool used to observe the quantum regulates the manifestation of the phenomenon. This, of course, most simply means that if asked whether

20 Bub, *Interpreting Quantum World*, 212–22.
21 Becker, *What Is Real?*, 68.

light is a particle (photon) or wave, the tool used dictates how it behaves. By the late 1920's, the HUP is fully being explored (as a claim that a single photon used to illuminate a particle is enough to alter position or momentum and therefore render measurement of both impossible) to blossom as the orthodox Copenhagen Interpretation, which stated that any description of the quantum is undefined before measurement and has instead a potential series of values with corresponding probabilities.[22]

Famously, in October 1927 at the Solvay Conference, Einstein rejects this as being of heuristic value. What Einstein meant by this was that quantum mechanics was an incomplete theory, inadequate to the task of describing reality as it really is at itself. As Jeffery Bub notes, it was not the indeterminism of the quantum world that he objected to but "[Einstein] objected to taking such theories as anything more than predictive instruments that would be ultimately replaced by a complete explanatory theory."[23] In short, Einstein worried that there was some kind of boundary on which one side was the HUP and quantum mechanics and on the other side was our world.

This, of course, as is well-documented, set off a series of conversations and thought experiments by Einstein to the Copenhagen theorists, most notably Bohr (with disputed clarity). Two are of note and bearing to this work as they still hold power in undermining quantum theory as being impossible and admitting the possibility of something more ordered (hidden variable that is for the theologian a kind of divine will in ordering probabilities) in the mysteries therein. In 1930, Einstein proposed his "box of light" thought project, which challenged the nondeterminism of the HUP (as one can never know simultaneously momentum and position, one can never determine where a particle will be in a future spacetime). Einstein's box emits a single photon after being weighed, is re-weighed, and according to Einstein's special theory of relativity, this would give both energy change and time of emission. Thus, it would be denying the HUP associated with the emerging Copenhagen theory. However, beginning to suggest the completeness of the system under study, Bohr responded by arguing that this violates Einstein's own general relativity as it predicts that photonic release alters the gravitation field, and this

22 There are differences even in the founders on the Copenhagen Interpretation as it is understood. Heisenberg felt that Bohr's interpretation was a confusion of potential states rather than actual states (i.e., when measured), which then fell to a more classical model. What Heisenberg thought was the difference between mathematics (including matrices) and reality as subjectively interpreted is the key idea so that the mathematics described the whole but measurement itself caused a state collapse, and this was, more or less, classical. See, Jonathan Allday, *Quantum Reality: Theory and Philosophy* (Boca Raton, FL: CRC Press, 2009), pp. 411–33.

23 Bub, *Interpreting Quantum World*, 4.

in turn undermines a prediction of time as space and time are linked. What Bohr pointed out would be that in quantum mechanics, as in the core-field theories emerging in the 1950s suggest, the relativity is fully accounted, and even necessary. Bohr, at least to this author, answers Einstein by referring to Einstein's own challenge to incorporate no boundary of the very small and the very large (general relativity). How this will look involves some deeper thinking, but it is not mysterious and spooky. The second challenge is more complex, the Einstein, Podolsky and Rosen (EPR) paradox. Here two entangled single-state particles (the total of their spin is zero) are sent in opposite directions, and on measurement of one, the other must have the opposite spin to make the system a single state. In classical physics, this would mean that one had one spin always while the other always had its opposite. But, according to the HUP, on measurement the wave function collapses to a fixed state of spin, which must at the same time collapse the other particle. This would require "action at a distance," "spooky action" or faster-than-light communication, which is impossible according to Einstein's special relativity. Thus, for Einstein, Boris Podolsky and Nathan Rosen the experiment (if done) would require hidden local variables that quantum mechanics has not accounted for, which dictates the entangled particle realities. In other words, not *probabilities* but realities are not yet understood. Quantum theory was a special case of a more general lawfulness, and the universe was not the probability of Eigenstates but borne out of a universal lawful purposed mind. Bohr's response (which sums the present field today) was unsatisfactory to Einstein, mainly that the entire system of particles and observer/measurer is a single system (even given distance), which does not contradict the theory. Eventually this would be understood more fully as *decoherence*, which mathematically describes and predicts quantum field theory in the classical world. It seemed that Einstein was carrying the day until John Bell's 1964 theorem proved the Copenhagen hypothesis by initially agreeing with the EPR paradox but also suggesting that if experiment proved otherwise, then the Copenhagen hypothesis was the truer representation of reality, despite its complexity.

Bell's theorem is again highly complex and technical. It is both mathematical and philosophical in nature but the barest bones as follows: Using entangled photons instead of electrons, and detecting polarization rather than spin, Bell theorized that Einstein, Podolsky and Rosen were correct—the quantum world had interactions that were local and hidden—and his famous inequality principle proved the Bohr system (in which it was possible to have faster-than-light interactions across great distances) was incorrect mathematically if one assumed said locality (that the photons could influence each other) itself. Further, the mathematics of the quantum mechanics of the Copenhagen school seemed to violate his inequality (as argued by Einstein) and therefore was false.[24] Critically, however, if observation proved otherwise, then the assumptions of Einstein, Podolsky and Rosen were incorrect. Bell introduced a devastating challenge—either the mathematics were incorrect, or the theory was incorrect as postulated by the Copenhagen group. But if experiment demonstrated that it was true, then a return to the mathematics and basic assumptions was warranted. This caused two movements in the field.

The first pre-dated Bell but took on steam with his work, to use deBroglie wave theory as a guide (see further in the chapter) and the second to set up experiments to see whether Bell's theorem was correct. Experiments in the 1970s and 1980s would prove Bell's inequality theorem (and more strongly the *Tsirelsoninequality*) to be violated and instead suggest that the quantum mathematics were better predictors of entanglement than the classical picture suggested by the theorem. The early work of the quantum theorists, despite the intuitive incompleteness to the classical world of physics, was proving to be more powerful *in experiment* than not, and the thick description of that world was increasingly allowing mathematic abstractions to demonstrate, not the bizarre mystery therein, but the wonderful conformity to our world.

What does this all mean? Was Einstein correct—that the issue with the quantum is that it has no bearing to the real as we experience it,[25] and that the key to understand it is to overcome the deficit of understanding of the Real as in the quantum world? Or is the Copenhagen position correct—that the behavior—physics of the quantum is so vastly different due to our position in the universe at this state that we cannot use naïve realism to explore it, but instead must use mathematics and probability? Does the universe, especially the very small or very fast (speed of light), have a "counter-factual definiteness" that breaches common sense understanding and instead is gestured by complex, often seemingly guessed in variable or constants, mathematics? Is this the realm of the supranatural? Is this the place of God? Or in our case, consciousness or the soul emerging from some other reality beyond the natural laws in the same way that Einstein believed there to be local but hidden variables in quantum physics, of which quantum theory just gestures obliquely toward? Do we have wrong measuring tools, and instead,

---

24 For a short mathematical history of the theorem, see Gennaro Auletta, Mauro Fortunato and Giorgio Parisi , *Quantum Mechanics* (Cambridge: Cambridge University Press, 2009), pp. 567–627.

25 A real problem for the decoherence model is the consciousness of the observer as observation occurs, which is adding "multiple worlds" to the reality of an event by its own reality. Any information gathering system (IGU) is itself in the act of observation of an alternate wave function that can, by definition, not be both self-causing and causative of the decoherence of the event. See Bub, *Interpreting Quantum World*, 224–32.

as for Polkinghorne, should look to the toolbox of theology to understand consciousness as these fit better with the ideas we wish to explore in terms of consciousness, ethical minds who will be joined to God in the afterlife? All of the above problems are debated in the infamous boundary of quantum mechanics and classical physics. In the classical world, there was a realism that accurately matched the world of sense to the world of mathematics in a lawful manner. In the quantum, those who asserted it was reality were "naïve realists" (e.g., the awful logical positivists of Mach and Vienna) who confused theory with experiment and certainly breeched the strict divide by not understanding the role of the observer in the experiment. Understanding is not prediction, and prediction is not understanding per se. It is theory, and to confuse them is to confuse everything.[26] And do we not have the pantheon of Greats who, themselves, could not help but use quantum mechanics as a metaphysical trap door, whether Einstein or Schrödinger?[27] Do we conclude with John Bell that quantum mechanics is FAPP and the only solution is to give it its own logic, probability, algebra or many worlds? The first three are opaque, and never really solve the experimental realism aspect and the last is the best candidate.

But there rose another candidate, which again vexed the great Einstein as "being too cheap," and which returned to de Broglie's work on wavefronts. David Bohm (influencing process theologians later) in the 1950s redeveloped de Broglie to allow "hidden variables" in the wavefront but not ones that which violated special relativity. His work reconsidered Schrödinger's wavefront as an objectively real event that worked as a field into which real particles followed the wave function in spacetime. Each particle in the wavefront has quantum potential—analogous to potential energy as momentum—which is described by the totality of the wavefront but on interaction with another particle is influenced to collapse the wave function into something that is classical in nature—predictive and determined. In the wavefront, however, all particle motions are predetermined—contained within the initial conditions of the wave function until measured. The wave function determines—as spooky action or a hidden variable—what the physical shape of the particles is in our world. What is hidden is the particle's real position until measured, the entirety of the particle's momentum is "nonlocal" in that it remains independent of measurement and on measurement is instantaneously manifest. Bohm's work, and its "underground physics/metaphysics against the "shut up and

26 David Wick, *The Infamous Boundary: Seven Decades of Controversy in Quantum Physics* (Boston, MA: Birkhauser, 1995), pp. 153–55.

27 See Wick's amusing account of Schrödinger's work in his *What Is Life?*, in which the great physicist resorts to a vitalist account of DNA as having some life force inherent. Wick, *The Infamous Boundary*, 163.

calculate" world of military applied and industrial physics post–Manhattan Project,[28] would continue to point physicists out of the hole that Einstein felt quantum mechanics led toward, which is to place the observer into a position of determining reality, or at least confusing theory with reality. And where there is confusion, perhaps God can show up.

This brief survey of quantum mechanics, including the absurd possibility that it is the mind of an observer that causes state collapse, demonstrates two deep possibilities: that the proposal of Wigner (and later Roger Penrose) of having a mind influencing state collapse is (1) scientifically untrue as quantum mechanics (extending to field theory) remains the best predictor of what experiment demonstrates is confusingly and (2) theologically problematic as granting it as an interpretation of quantum mechanics is not a refuge for God (or mind/soul). The theological problem was best given by John Wheeler's application of the participatory universe in which present observers (God but also humans and by extension all observers, including animals and possibly plants) influence the distant past, including the post–Big Bang cosmos. Here we have two basic theological problems: (1) all observers are seemingly central, including future ones, in the becoming of the universe and the sustaining of it, which raises the question of freedom itself (e.g., the past, including my own, is influenced by a future observer of the universe), and (2) God is reduced to one agent, perhaps the most significant one, in the plentitude of future participants in the universe. This is hardly the theistic God of the Judeo-Christian tradition.[29] It is, as in the neuroscience challenge, not a definition of the soul that fits with any of the core ideas carried in the theistic traditions. It is an education to a different reality of humans in nature.

Understanding Symmetry

As indicated above, the issue in quantum field theory is that it is the assumption, FAPP, that decoherence somehow allows the physicist to peer backwards in time to the nonclassical world of quantum mechanics and its various superpositions and supersymmetry by the observation of the collapsed wave functions in our macro-world of atoms, galaxies and black holes. This decoherence, further, has an infinite density of entangled quantum states, which manifest locally always the same—our classical physical world—but does so in a different space and time.[30] The quantum world is "entangled," while our macroworld

28 Becker, *What Is Real?*, 193–218.

29 Allday, *Quantum Reality*, 428–34.

30 See E. Joos et al., *Decoherence and the Appearance of a Classical World in Quantum Theory* (New York: Springer, 2003), pp. 1–41.

is not and there is some kind of "Heisenberg cut" in our low-energy, low-wavelength universe of the possible entangled states of the possibilities of the great moving wavefront function of all reality. Some new ideas are now introduced. In quantum physics, symmetry is all the possible states that any fundamental force (or attribute or constant) can exist within the constraints of its interaction with other forces—it is notional in terms of mathematics—whereas most things in reality persist in broken symmetry, which is the specific interaction(s) within the possible manifestations. Thus, electromagnetism has mirror (left and right) symmetry, which allows the possibility of DNA as manifesting in one way (right helix) but it could be in both modes of left and right without violating any law known. That it is right, and all life is right, is merely a stability issue rather than a condition or that Dirac's electron being negative is merely a condition of stability rather than possibility. The system itself chooses to be stable—manifest—out of the possible states it could be and as long as these do not violate any laws, it really doesn't matter in the exchange of stability in dis-symmetry. While many states are possible in open symmetry, interaction dictates broken symmetry and is stable. Further, it is only at very high or very cold temperatures (quantum vacuum) that symmetry is able readily to reassert new and independent symmetries.

What we do know classically is that symmetry is influenced by energy/heat, as evidenced by phase transition in liquid or heating a magnet beyond the Curie temperature. In both cases, what is observed is an exchange of stability for idealized symmetry as it really doesn't matter what kind of electromagnetic field is generated by any monopole and as energy is introduced, the monopoles can take all possible forms that are in essence no longer unified, and a magnet loses its field.

Our universe could be a "kind of monopole magnet" in which the initial heat of the Big Bang creates a random set of possible symmetries that, as they cool, break into stable forms, and if there is enough expansion between the symmetries, could be considered universes in themselves with their own broken symmetry as stable constants. Thus, constants could be low-energy effective laws in a single domain.[31] We could, further, understand the interactions of those broken symmetries as gauged symmetries—ones that influence the others—so that an entire group theory of interactions makes sense for one domain. Further, it could also be possible that what was once thought as a fundamental constant could be blended with other constants at sufficiently high energy so that the broken accidental manifestation is really a reality of the larger symmetry (e.g., weak force with electromagnetism in

31 Davis, *Goldilocks*, 163.

the Glashow–Salam–Weinberg theory) despite that perfectly predictable laws worked at the lower stable temperatures.

This is of great importance in terms of the Higgs field, as a zero Higgs field is symmetric but unstable and as it cools, it exchanges to a non-zero value and particles resultantly assume masses that then in turn effect all fundamental interactions with each other. This is of importance as the temperatures predicted for this are so high that they exist before 10^{-36} seconds after the Big Bang. We live, even in the quantum, in this low-energy stable universe.

Adding further complications by extension is that string/M-theory considers interaction beyond mere 3D standard model so that the actual folding of spacetime itself becomes an added player in selecting the broken symmetry of constants. Reality is then a kind of landscape of low-energy universes populating and interacting as they expand, bump and contract. The nucleation is quantum, below the 3-D universe. But the point is simple—once broken, it is stable and predictable. This leads us to two pragmatic and resultant theological or philosophical conclusions:

(1) Local knowledge: We inhabit a local universe within a much large singular universe (imagine a vibrating, crumpled bit of bubblewrap—the universe is the entire bubblewrap, and we are but one node in the crumbled mass that is vibrating uniquely depending on location)—there may be so-called quantum laws in play, but these are largely (a) unknowable through classical models at present and (b) unprovable by classical models. In short, we can only do the math for here. This is attractive to theists for obvious reasons as it possibly permits a "God of the gaps" theology extending beyond realism or physicalism.[32] This is the heart of the debate between Einstein and Bohr on the Copenhagen interpretation, and the place of consideration for those who wish to answer the "measurement" problem and accuse QFT as being too presumptive with its need for superpositioned kinematical models and introduction of dimensions such as Hilbert space, nucleation or other such models. But as physicists since Dirac have argued, and observation/prediction has consistently demonstrated, the decoherence of superposition is mathematically viable and experiment bears out repeated predictions of such collapsed kinematic systems.[33] It does not mean, however, that the measurement problem is solved and philosophically speaking, the Einstein trajectory is correct on his claim of anti-realism of quantum mechanics.

(2) Infinite or many-world universes: We inhabit one branch of a singular universal wave, but which is always in collapse depending on observation or measurement, and with each collapse, a new universe is created. At

32 Carroll, *Big Picture*, 212.

33 Joos, *Decoherence*, 39–40.

the quantum, it is possible to think that every interaction is a possible new reality and thus there are infinite possibilities. There are multiverses, each derived from a set of unknowable quantum laws that fixed into predictable local constants. This is a postulation that cannot be denied and is verified by experimentation, and forms the basis of superposition, supersymmetry and QFT.

The attraction of (2) is great for theologians and philosophers. Core field theory is really just (1) above but with the reverse claim of proof placed on religious minds as in the work of Sean Carroll. But the many-world universe claim is a candidate for theology (and especially process theology), or, at least, consciousness and souls.

Many Worlds Interpretation of Quantum Theory

From the outset, there is not one generally accepted version of the theory, which is the earnest attempt to bring general relativity and quantum mechanics into one core theory and to solve the initial (and persistent) problems of the Copenhagen position and its early experimental proofs. But what all agree on is the logic of the theory, namely (as formulated by Hugh Everett III) that the inclusion of observers must be accounted for in any physical system within a theory. The reason for this is quite simple (although it could apply to any information gathering unit (IGU)—human brains/minds are the things we use to observe reality, and this is a complex chain of events from vision to "thinking," each of which is a physical event according to quantum mechanics, and therefore each "pushes" back the arrow[34]of time to causing decoherence of superpositioned particle symmetries that are plucked out of the wave function as a "branch." Erich Joos, picking up on the work of Roger Penrose, treats the mind as a quantum computer, with each neuron firing as not only having millions of potassium and calcium ions involved but also needing for each to have a decoherence timescale for the singular event of one neuron, let alone thousands, firing or not. We, he notes, measure neurons firing in milliseconds, whereas the decoherence of a single ion occurs in a zeptosecond. Each collapse, in turn, separates a branch of the events that follow. The free evolution of the wave function is clearly scattered into millions upon millions of branches. Does this mean that the mind, let alone the singular ionic collision, creates reality? Are the multiverse branches in the act of observation raised by Zeno's postulate (on the flight of an arrow in which it is literally frozen in time and space such

34 Joos, *Decoherence*, 108.

that it cannot be said, logically, to be in flight at any one point of its trajectory) such that you have an infinite number of events without continuity? There is no multiverse, just short infinitely small events out of sequence like film stripes cut and randomly scattered. Is there a quantum Zeno effect? Is the multiverse of decohered particles an endless series of independent realities? Or are there, as Joos and physicists after Dirac postulate, mathematical models that allow for such calculus? The calculus of such seemingly independent Zeno effects is the consistent histories of the wave function.[35] This allows QFT to work the arrow of time, in duration and direction, backwards from the consistent scientifically observable and predicted macrouniverse to the quantum.

Steven Weinberg's lecture for the 1986 Dirac Lectures is helpful here. In his lecture, he deals with the "reduction" of quantum mechanics as a kind of system in which the theorist takes the obvious state vectors of elementary particles and expands them to larger systems including this author, the planet and universe.[36] This, he notes, is both deterministic in evolution and indeterministic until measured. This is exactly Zeno's paradox on the flight of an arrow applied if we begin to spin the archer in different directions (perhaps standing on a spinning planet hurling through space). The arrow is indetermined in the sense that there are many places it could be in this larger frame of reference (but can be understood as a state vector) but at any one moment of time, it collapses to "here" and is determined. The discussion of this evolution of an arrow, and the arrow can also be spinning (having an electrical charge, etc., as it moves and interacts with other things such as air molecules and gravity, all the way to the individual electrons and fundamental particles) is a discussion of a summation of state vectors. The job of the physicist is then to relate the various state vectors into a larger symmetry. Classical mechanics did this, and classical quantum mechanics also did this in individual cases of simplified state vectors, and the job is now to find a way to mathematically bring them into discussion with each other. Weinberg's lecture is a defense of symmetries of QFT as a way of looking at elementary particles and reconciling general relativity with some real experimental force. Symmetries, then, are the mathematical manner by which the problems of the Copenhagen Interpretation are solved, and the then sterling candidate—string theory—is merely a mode of mathematically discussing these various symmetries, including gravity.[37] The point is simple:

35 For a discussion on quantum tunneling and quantum evolution using Hamiltonian equations, see James Bayfield, *Quantum Evolution: An Introduction to Time-Dependent Quantum Mechanics* (New York: Wiley, 1999).

36 Steven Weinberg, "Towards the Final Laws of Physics," in *Elementary Particles and the Laws of Physics* (Cambridge: Cambridge University Press, 1987).

37 Weinberg, "Final Laws," 95–107.

> This is the punch line, the symmetries determine the action. This action, this form of the dynamics, is the only one consistent with these symmetries. In other words, there is nothing else you could have added. There are no other possible terms consistent with these symmetries [...] that the symmetries of the theory have completely determined the structure of the dynamics, i.e., have completely determined the quantity that produces the rate of change of the state vector with time. [...] This theory has the smell of inevitability about it.[38]

The key point, however, counter to the metaphysics of Becker and the Copenhagen dissenters is not that this is a strange alien world of mystery but that it is exactly the world we measure, interact with and can make predictions about—just as in both classical and classical quantum mechanics. Weinberg notes that the great Dirac's theory of the electron—and Dirac dissented from Copenhagen sharply—was an early example of unification of quantum mechanics and special relativity so successful that it predicted antimatter but also fails in the discovery of other elementary particles that do not share the electron spin of one-half such as W, Z and then predicted Higgs bosons. Those symmetries had to be included in the new theories. In the case of the W and Z bosons, they had been predicted (by Pauli in 1934) and observed (1973 and 1984), and in the case of the Higgs only recently observed. Weinberg's lecture, at the time of the first foray into string theory as a form of QFT, reiterates the lack of metaphysics in the mathematics in that it is the real observable world of the classical picture that is tested in the mathematics.

This has the sharp differentiation from Wigner's claim that mind or consciousness (the observer) is separated from the material universe being described by any quantum collapse, or in fact collapses the quantum superposition. The observer in physics-speak is simply a vector in Hilbert space that has a unique ability to describe, as memory, any moment of the wave function collapse as part of it, and not as causative. Thus, the famous measurement problem of quantum mechanics is recognized as a feature of human subjectivity just as a computer—or any IGU including a dust mote—may be made "aware" of an event and use it to predict certain future events but not cause those events by the measurement. It is a part of the system itself, self-aware of any event, recording the event and then deducing the event but not causing the event. The event, say a computer measuring a hit to a website, did not cause the person surfing to land on it, but can predict based on the hit something of the identity of the surfer.

---

38 Weinberg, "Final Laws," 103.

For Everett, and necessary to avoid infinite Zeno branches of events in the act of recognizing an event subjectively (e.g., I measure, I am aware that I measured, I reflect/relate on that measurement which is a new measurement of sorts and so on and in doing so am experiencing a collapse of superposition of the quantum each time at increasingly slender intervals of time), it is not so much that there is a new universe being created by each act of measurement or awareness but rather that there is one universe (ours as distant, cold and so forth) that has many viewpoints—and if we had a powerful enough computer or IGU, could "see" all viewpoints behave according to the one singular core theory at every point of spacetime.[39] More simply, in Hilbert space, observers have an infinite number of permissible coordinate frames according to relativity so that any one observer frame "frozen" is just one of an infinite number of position coordinates in our expanding, slowing universe. We do not cause the collapse, just notice it as we are in it. It moves with a psycho-physical parallelism, we are physical agents in a massive quantum wave, but it does not mean psychophysical reality is caused by us.

The genius of Everett, and near theorists such as Bryce DeWitt, is that they wish to avoid the very thing that many quantum theorists such as Wigner and objectors such as Einstein felt was in the bones of the science—namely, that the quantum was ontologically different from classical reality and that observation collapsed a waveform, which naturally led to the thought that a mouse observing itself created an independent world at each moment. Pushing this to its logical ends, observation of distant stars (our past) creates our future or the smallest firing of a neurochemical in the brain splinters reality, given that simple sensory sensation involves millions of such interactions, it become blindingly complicated. Some objectors, using both philosophy and mathematics, argue that the many-world theory with its de Broglie–Bohm pilot wave creates empty waveforms with decoherence or collapse, and that this means not merely many worlds but many universes, some empty (just being potential) and some collapsed. Of course, what collapses them is the presence of minds, and this is solipsistic—a mind not only creates itself but its universe out of many empty ones in potential.[40] But the response to this is simple, why bother with this idea? What is in play is a rather elegant concept—that Einstein himself would have approved of, and did with his EPR sufficiency criterion—namely, that if a theory (here de Broglie–Bohm pilot wave function of the core

39 David Bohm and Basil Hiley, *The Undivided Universe: An Ontological Interpretation of Quantum Theory* (New York: Routledge, 1993), pp. 296–99.

40 See Antony Valentini, "Pilot-Wave Theory: Many Worlds in Denial?," in *Many Worlds? Everett, Quantum Theory and Reality*, S. Simons, J. Barrett et al. (Oxford: Oxford University Press, 2010), pp. 476–509.

theory) predicts a measurement will yield a certain value and it is found to be true, then the measurement is revealing a property of the system inherently. The fuzziness of all the Eigenstates of the wave function in superposition is certainly questionable, but the theory as a tool for explaining is not. Why the metaphysical then?

While relativity demands strict continuity, causality and locality, quantum mechanics seems to argue the opposite and the quest for an implicate order which includes observers who are placed into a waveform that allows for all causalities, continuities and nonlocality via entanglement, many-worlds interpretation is a way in which there can be a marriage, and a marriage that works in our universe. The many-worlds interpretation uses the idea that wave function of implicit order (enfoldment) when it collapses (unenfoldment) returns to a new wavefront within the original, but not superseding it. It all works mathematically with Feynmann graphs and Huygen's constructions that elude most but the most advanced mathematicians. But is neatly summarized as "giving history a real content as objective and independent of observers"[41] and to treat our universe—one of the multiverses—as a particular aspect of a deeper super implicate order—that of the quantum world. CQFT is not, then, metaphysics but physics. Consciousness is not the cause of CQFT, but CQFT is the cause of consciousness. Or in terms of theology, souls are not the cause of CQFT, but souls may be the result of CQFT.

Re-Understanding Quantum Mechanics: String Theory and Supersymmetry Redux

As is well known, the origin of quantum mechanics came from the wave–particle duality of Planck's double slit experiment in which he proved that light is wave and particle, or wave and photon, and that it was impossible to predict which was the true behavior (e.g., particle or wave) of light. Further work by Heisenberg demonstrated that a quantum object cannot be ascribed a total set of attributes at any point of time because fixing one attribute (mass, spin, position, speed or momentum) impacts the others so that it is impossible to fix the total set until accounted for as a set and fixed by Planck's constant (as the scale constant) using Dirac's work. In lay terms, imagine all possibilities—draw a line through them (it is much more complex but see Feynmann on this) and this is what you can say "will happen." The "drawing a line" is fixing the set, a mathematical shorthand that takes all possibilities and fixes them and this, surprisingly, is what we "see" really, so that light moves in a straight line

41 Bohm and Hiley, *Undivided Universe*, 367.

at fixed speed and so forth. This is classically quantum mechanics, and QED, but CQFT theory needed more (as above).

Running current to the work of CQFT, dealing with the issues of experiment led to the conclusion in the 1950s and 60s that current field theories did not agree to observable data, and with interactions of quarks that did not become stronger as energies and momentums grew but were weaker, physicists needed a new way to interpret the data of observation. These interactions—initially grouped as Yang–Mills theories on non–Abelian gauge theories—were studied and predicted new particles such as the Higgs boson and these weak interactions became the foundation of the Glashow-Weinberg-Salam (GWS) electroweak standard model. But to account for this, the field theories had to include a larger gauging—more dimensions to build on than in our ordinary world. Thus, CQFT was working simultaneously on wavefront ideas and then adding in what was observed and seeing there was demand for explanation that took us "out" of our usual 4D world (including spacetime). Initially, this was not much of a problem, and irregularities in quantum numbers were called "color." However, this quark model as later experimentation in hadron colliders would verify was both massively (pun intended) successful and massively problematic. It was successful in that the model predicted colors and vectors therein as mass, energy and spin that were stable and could predict particles under the quarks to include mesons and charm quarks as well as the Higgs boson. But it was problematic in that it didn't fit well with general relativity (again), could not account for the nonzero cosmological constant and things in abundance such as baryons, dark matter and the basic metric structure of the universe—and one more little thing—gravity.[42] How could these particles, reality and observation fit into the new physics?

Physicists knew that quantum interactions are "virtual" in that they are possible in all possibilities as a field in perfect symmetry but as the number of interactions occurs in superposition, the field collapses and becomes singular. Quantum interactors include quarks (strong force) and leptons (no strong force), and then virtual things such as photons, gravitons, bosons and gluons, and each have unique spins, mass and charge. At the quantum level, these interactive particles interact with predicted realities such as the Higgs field and in doing so begin to acquire and lose mass so to blend forces as in the GSW theory of electroweak force confirmed at Cern. This is a blending of the standard model with quantum chromodynamics (QCD) and usually involves high energy. Quantum mechanics is both a place of extraordinary conception, driven by what is indeed observed, and yields despite its conceptual difficulty

42 See Michael Dine, *Supersymmetry and String Theory* (Cambridge: Cambridge University Press, 2007), pp. 1–135.

a consistent model of reality or IBE. It does not, nor ever has outside of science fiction, required God or the Q continuum of Star Trek. But there were questions to be answered, parts of the puzzle and reality that were not accounted for in the theories. To answer this, physicists begin to work in the seemingly bizarre multidimensional world of string theory, and the theologians, as before, flocked to the bizarre as evidence for both divine activity and the paucity of science itself. But, again, is this really what the physics demands? Is string theory metaphysics, and how did it come to be a possible candidate for the natural physical explanation of reality?

The primary problem is gravity because gravity works not as a force but as a warp of spacetime itself (Einstein's relativity). Gravitation is brought in by string or the more generally accepted M-theory version and pays attention to the issue of missing antimatter from the origin of the universe. Gravity is the last unknown force and the last game in town for the theorists to work from.[43]

The starting point for string theory is that there is a universe, and this means that antimatter and matter could not be balanced out as there would be nothing at all except gamma radiation created after annihilation, by asking questions as to how the universe must have quickly quarantined the two matters on a very large scale to make it impossible to interact. So how did 10^{50} tons of matter get created without the same amount of antimatter? But as important, that there must have been a universe—dimension—in which the various forces did indeed work as a singular field with an equivalent field equality such as we are used to following a Coleman–Mandula theorem. This "place"—where all of the forces we know combine to be "one field" occurs at extraordinarily high energies (above 1 TeV) and to include gravity, is what the mathematics describe.[44] This place—the supersymmetric unified field—is not our universe/world/dimension—we instead live in a symmetry that is broken, which explains the problems of the standard model. Something must have broken the symmetry—but what can this be? Using the standard model, the only unaccounted force is gravity and thus string theory is an attempt to understand gravity and uses, and this is important, the known ideas of quantum mechanics to introduce a particle called a graviton that exists only at high energies in our universe, but nonetheless exist. Gravitons are the key to string theory and are particles from other energy dimensions that interact with ours, and may in fact, cause our physical world.

String theory comes out of the success of the standard model of particle physics, the need to find a field that accounts for the unification of three forces plus gravity, and then other observed problems in cosmology and particle

43 Wick, *Infamous Boundary*, 186.
44 Dine, *Supersymmetry*, 139–55.

physics, including (1) moving the energies to the high levels of origin (basically we are good to 100 GeV) and then we encounter issues with our constants, including the cosmological constant that is fantastic in its utility suggesting dark matter; (2) baryon to photon ratios; (3) the lack of homogeneity in the universe we observe, and its flatness topologically and the lack of predicted things such as magnetic monopoles.[45]

Overcoming Pauli exclusion (no particle of the same quantum state can occupy the same state at the same time) was to conceive of supersymmetry as to explain gravitons as possessing a predicted spin of 2 (even integral particles are bosons with half-integer particles are fermions and both make matter but Pauli exclusion problematizes how they can coexist unless one begins to rotate the spin axes of fermions so that together they as matter are always 0) and thus the hunt for "s" partners occurred or gravitons that are in different spatial orientations. Gravity possesses some "s" partner, whose geometric interaction is an example of supersymmetry. Because gravity also has energy, it is impossible to describe gravitons, unlike other "particles" whose Heisenberg fields can be accounted by "infinity mathematics," in either classical or Copenhagen formulations. They must be reconceived, and rather than a particle, they are thought of as strings or loops that subsist in different symmetries or vibrations at once. What we "see" in the loop is the graviton, but even here we have not seen it as total as the energy is not high enough.

Gravitons as strings are capable of different vibrations (even in the same string) to interact with supersymmetric partners and have the needed "s" partners, which are demanded by the physical observation of the universe we inhabit. Further complicating string theory is the size of the strings (a hundred billion, billion would not cover a nucleus) and the fact that gravity warps spacetime so that compactification occurs, and they may exist in other dimensions of spacetime. M-theory also notes that the strings of gravity are not in fact strings but membranes or straws that only looked like loops, but in fact where 10 dimensions of quantum interaction and M-theory offer mathematical descriptions of how those interactions of the four fundamental forces may look. The point of this is that M-theory is odd, but it is a mathematical model—it assumes that there is an order or constant in nature that is not a sliding mystery. It may not be a GUT, but it doesn't deny the possibility that there is a fundamental law(s). Michael Dine takes the complexity of string theory and reduces it neatly as a theory that takes the standard model, general relativity and our best inference to the above phenomenon as "not postulated [...] [but] they are inevitable."[46]

45 Dine, *Supersymmetry*, 269–72.
46 Dine, *Supersymmetry*, 307.

String theory is certainly bizarre, conceptually complex and assumes much in terms of the mathematics and energies therein, but it is also a way to explain the observable universe we inhabit. It may be incorrect, premature in formulation, but it is open to inquiry and to be fact checked by observation in this universe. Returning to Weinberg's claim of its smell of inevitability—it both has predictions and applications and is mathematically beautiful—meaning making sense of what we have measured, observed in our cold, slow field. Our universe is the universe, or at least one suburb. Exploring our suburb using the tools we have—call that science—answers our fundamental questions. To ask the nature of field symmetries is to ask about our universe and it may be an incomplete heurism, but it is a valid question. This may smack of positivism or reductionism, but it works and the gap between the world of experience and the world of theory narrows. It certainly, as in the case of neuroscience, does not mean that consciousness collapses the world into a fracture of zeptosecond empty and full universes.

So why does this all matter? The problem is that the confusion is a conceptual one, not a mathematical one—the mathematics posit an inference to best explanation (see next chapter) or an IBE, then we look for this and more often, if we look correctly (say black holes as in the work of Hawking, we see the mathematical predictions in reality). Quantum mechanics/chromodynamics, QFT or even string theory is not at all a mystery in terms of what can be predicted by the mathematics moving from known to explanation of phenomena observed, it is a mystery in terms of understanding those processes completely today or tomorrow, but this is not a "God of the gaps" theory. It is expected that, and even unfortunate as in the discovery of the Higgs boson, predicted particles such as the Higgs boson be seen. It is conceptually difficult, not conceptually impossible and the driver is observable data. It challenges our understanding of causality (bottom up, arrow of time, etc.) but it is not causality outside of the system—it is broken symmetry and is (hoped for) predictable even in the quantum vacuum.

The Metaphysical Implications

For most physicist-philosophers, the implications of the quantum revolution—whether in QED, QFT or string theory—leads to two broad areas, excluding the Wigner hypothesis of consciousness causing reality (dealt with above). These are:

A. We are a simulation universe—perhaps as Nick Bostrom argues,[47] we are a computer simulation (Turing Machine) of another universe and

47 Nick Bostrom, *Superintelligence: Paths, Dangers and Strategies* (Oxford: Oxford University Press, 2014).

merely players in some Star Trek super-alien race computer game. Our universe is simply the selection of their preconditions for it, including our movements within and ideas of freedom. However, as attractive as this may seem, it is impossible to know whether we truly are and thus this is a moot reflection. The players cannot ever know that they are a game.

B. Death of theistic God:[48] The universe cannot have design, no teleology.[49] The universe cannot have an arrow of time as we consider it with a purpose.[50] If we have a God, it is a demiurge as equally trapped as we are in the broken symmetry of the known universe.

The simple point is this: The above are intrinsically hostile to any traditional account of the theistic traditions found in the Abrahamic traditions with their arrow of time and teleology, and the account of the uniqueness of this universe. Einstein, for example, was certainly not impressed with the classical theistic God of his Jewish ancestry, more oriented to an impersonal "pilot wave" God of Nature as found classically in Spinoza.[51] Mystery here is largely philosophical, hostile, and contrary to the birthing tradition of Judaism and Christianity, and certainly does not assume that the mathematics is arbitrary—it is merely finding the correct laws for what broken symmetry is expressed. It still, at least in theory, allows for a Grand Unified Theory (GUT) that will have predictive force. To paraphrase the Marquis de Laplace or Lawrence Krauss on God—we do not have need of that hypothesis. But we have not explored the mystery of the quantum, even if stable as the quantum world is one without time as we conceive it according to Einstein. Time, as Sean Carroll notes, in quantum physics "emerges as course grained level of description" but the universe(s) is timeless.[52] The universe does not, and this includes humans, come out of nothing—ex nihilo—because our universe comes from another multiverse and time is simply our spacetime manifestation of the core theory. It came into existence, but according to an understandable lawfulness that quantum physics gestures. The multiverse is true, and our universe is one case of knowable constants in broken symmetry. The best explanation, in short, of the universe is not God, but physics. The only necessary being is existence, and this is best explained by physics.[53] Add to this the last piece of information—that

48 Davis, *Goldilocks*, 186.

49 Davis, *Goldilocks*, 198.

50 Davis, *Goldilocks*, 200.

51 Jim Baggot, *The Meaning of Quantum Theory: A Guide for Students of Chemistry and Physics* (Oxford: Oxford University Press, 1992), p. 202.

52 Carroll, *Big Picture*, 198.

53 Carroll, *Big Picture*, 204.

the laws of physics are not, except for existence itself, a definition of "why," and one is left with an uncaring universe; one full of violence, most of which threatens the fragility of life as we have come to understand ourselves and not that of a caring, loving Creator—life seems despite the universe, and not because of it. Surely, the classical theist position must account for the brute fact cosmos we observe—one of uncounted galaxies disinterested in a little planet in the spiral arm of the Milky Way? Pushing the analogy back to the computer simulation, whose pleasure is the violence of the physical world within the "World of Warcraft"?

But, even admitting that origins may be conducive to belief in classical theism—that the universe began as the product out of nothing but divine love with the centrality of the human species in a Jewish tribe, Christian sect, or general humanity—we still have another more direct problem theologically. Namely, whether the body and soul of classical Judeo-Christian theism understanding that includes ideas of afterlife is compatible with the new physics and core theory? Eventually we come to rank substance dualism, or any other form of dualism, in terms of physicalism, and a physicalism determined by quantum physics and core theory. The physicist will argue that mind or soul is merely a way of talking about the behavior of certain collections of physical matter. There is nothing else in the core theory that is unaccounted for—no room for the gap of the soul. Nor does the property dualist fare better—matter is matter as understood in the core theory—it has only the properties of the core theory and there is nothing behind it but the physics. You would need to argue that somehow the core theory is incomplete, and the FAPP principle is insufficient. The modern mind is given two choices—theism or core theory—as explanations of "why" the universe is and why we are. Given the power FAPP of core theory, do we need to use God to explain the universe? The only place to do this is morality or value, and this project is left to be explored in the final chapter. Returning to our introductory chapter, core theory makes it improbable that (1) humans are uniquely created with divine purpose and (2) makes it improbable that humans are uniquely—due to soul extra—related to this Origin.

Finally, core theory makes the bedrock ideas of (3) afterlife and (4) account of divine activity in Torah or Christ untruthful because death is the end. We are, according to core theory, our material selves and when this is no longer—the atoms that arrange to be us—then we are done. The wavefront is collapsed and finished for us, fixed in a moment of time and no field, no particles can take "us" to another body, another time or another place. The core theory says we are things happening, and when that multiverse is ended—there is nothing after it. We are quantum fields—from our hair to toes, from our brains to our thoughts—and when they have decohered and end then that world is closed.

In this sense, we are constantly dying. But we do not transform. The end is the end.[54]

This deep foray into physics is simply to argue the point that emergence as consilience is not a random attribute of some capricious and unknown physical reality within our physical reality. If it were, then we end up with the demiurge God or a completely random universe of macrocausality and not the theistic God. Light, at the macro level, behaves quite predictably because at the quantum level the attributes selected in the collapse of the wave package make it such. The ghost in the machine of the physical is not necessary, and in fact, may be an issue that comes about from our understanding of nature that is itself our issue. What we have is the idea of talking about ourselves in a useful way. Theism was that candidate, and a very good way. But it was a fiction, a symbol and now we have something else. Physics, and then it extended into neuroscience as we are brain all the way down.

But then the question remains, can we live with this? Is this "brave new world" really a "Frankenstein" of master race mentality, eugenic amorality and dystopian? Isn't the real villain not science as truth, but science usurping morality—the last frontier of human uniqueness? This, then, makes us face our place in nature as naturally natural and why we might be prone to dismiss this, and why theology has been such a good companion to resist this brave new world but now may no longer be useful. Have our Stone Age souls finally found no appeal to the modern mind? Can we educate to reality and become nature realists, accepting we are brain all the way down? Is there a middle way between the physical world of neuroscience and core field theory and human description of reality? Which one is more helpful at present, and likely to continue to be more helpful? Finally, is helpful the same as true?

---

54 Carroll, *Big Picture*, 215–21.

Chapter 4

THREE GREAT ERRORS: DECONSTRUCTING NATURE, NATURALISM AND SCIENTISM: FICTIONS OR DELUSIONS?

A widely considered master narrative of the past century is of a steady disenchantment of the Judeo-Christian Platonic world and the rise of a soulless scientism of Aristotelian causality.[1] The story is as follows. When the sacred canopy—the place of God and morals—is eroded by naturalist, positivistic scientism (see Becker on this with respect to quantum mechanics) then what grinds out supranature is extrinsically hostile to humanity and indeed creation. All the villains of social Darwinism, technocratic Prometheanism and scientific reductionism manifest as philosophical naturalism[2] to deny morality, connectivity and certainly *conatus* or drive for moral reform, let alone connection with one another and the wider planet. Humanity is left to drone mentality, over and above nature and aimless in a cold distant nothingness.[3] Into this story, the enchanted or mystical is preferred, whether Real or not, because it pulls out of mere representation toward imagination, and in doing so, provides ethical will and energy.[4] This repeated and terrible story ends with the "human being [alone],"[5] and our more recent postmodern, postfoundationist, postracial and so forth, history as the humane desiring to hope and a return to the native or naïve, and a re-enchantment of human purpose, connection and destiny. Imagination of a better, unconnected to our historical foibles, future is the remit and there is almost always an appeal to a

1 Charles Taylor, *A Secular Age* (Boston, MA: Harvard Press, 2007).

2 Charles Taylor, *Philosophy and the Human Sciences*, Volume 2 (Cambridge: Cambridge University Press, 1985), p. 267.

3 Taylor, *Philosophy*, 268.

4 Peter Dews, *The Limits of Disenchantment: Essays on Contemporary European Philosophy* (New York: Verso, 1995), pp. 7–12.

5 Andrew O'Shea, *Selfhood and Sacrifice: Girard and Taylor on the Crisis of Modernity* (New York: Continuum, 2010), p. 209.

better nature in humankind. This is, of course, a prehistorical, presocial and completely natural human being uncorrupted by the positivistic history of the Western world and its most beloved child in science. But little attention is paid to this presocial natural human for whom education (both formal and cultural as a kind of ontology) can resurrect for a better future without inequalities of every stripe. For our purposes, this narrative is important. In particular, the thinking on "natural" and "nature" is imperative as both those terms seem to be Edenic or preferential to the context of culture we find ourselves in. While we may be the product of cultures—personal (family and immediate context) and social that corrupt our moral and self-orientations through traumas—there is also the presupposition of a natural self or person untainted by those winds of context. The trick is to live in that nature endorsing connected self to the planet, to life and to each other. This is an odd take on J. J. Rousseau's human born free but put into social chains. This is not only an antidote to culture as the location of human dysmorphia but also, and importantly, an antidote to the technocratic Prometheanism of social Darwinism and scientific reduction of all human uniqueness, including for our purposes the idea of the soul and the complex of ideas such as consciousness, identity and so forth. This is "nature skepticism."

The general result is a "nature-endorsing" philosophy/theology that firmly embeds human nature within Nature as quasi-religious duty or religious dimension in response to the "nature skeptic" who understands reality in terms of largely science. These two terms—nature endorsing and nature skeptic—are used by Kate Soper to discuss the history of the West's relationship to the idea of Nature. Think Rudolf Steiner interacting with near contemporary Charles Darwin on the role of spirit in evolution. The question then is whether this is true or useful, whether it is science or theology and whether it is possible to construct a human desire for humaneness independent of God (nature skeptic) or re-enchantment of a divine impulse (nature endorsing)?

Nature and the Natural Redux

I believe the way to understand the thesis of disenchantment is to understand modernity as a shift from theological Realism to semi-Realism and finally Naturalism. This is an inversion of the work of the consilience group dealt with in Chapter 2, who argue from Naturalism through semi-Realism to Realism in the work of consciousness. Both, however, claim a ghost in the machine against sterile positivistic scientism. The shift is from a description of reality and *mores* in the premodern world in which a transcendent God guarantees the Real, which also promises something transcendental within the human person accommodating to said Real. This can be in reason, intuition, a vague connection or the Romantic sense that what was transcendent is transferred to the natural world in its totality, and in which human beings are but one aspect of the total. This is a shift from divine ontology to human immanence and finally to world immanence.[6] By most accounts, this shifts from a reading of the book of God to a book of theological anthropology and finally to a book of Nature in which the human is central. Of course, the positivist sciences in this narrative most fully decenter and finally push out human uniqueness, and by extension, Nature's uniqueness, as *Wissenschaft* grow in power. Most of the attention, however, is given to the death of God and the subsequent double death of humanity with its loss of imagined space and empathetic morals. Both the world immanence group and the emergence consciousness group share a movement against science itself and have a privileged view of nature itself. This is what we should reflect on. What is natural about Nature? What is special about Nature? Is Nature, in other words, merely an adjunct to human nature (nature endorsing) or is it merely something out there in the world, inert or disinterested to humans (nature skeptical)? Of course, the latter position easily leads to a disinterest in interdependence, while the former seemingly gives precedence to it. But is this the case? Little attention is given to the role of the idea of "Nature" understood in this transfer from the supranatural to the natural, and the role of the human within it. I propose to argue that the idea of nature held by the disenchantment/consilience thesis is in fact as problematic and misleading as that which it (in modernity) wishes to replace. This result is a false dialectic that pits the metaphysical and the physical as oppositional. What this means, in part, for neuroscience and theology is that the physicalist position is gleaned to be unnatural and the metaphysically nuanced emergence theorist brings mystery, freedom and ethics back into the fore. But it is equally a problem for theology. It denies the very thing—transcendence—that the Abrahamic traditions wish to uphold and slips into a pantheism or Turing Machine. Our first question, then, is whether Nature can be metaphysical at all?

The philosopher Kate Soper asks this basic question on the necessary relationship of Nature as infused with deity to several significant theological movements that break with modernity in feminist and liberation theologies and who share a Nature metaphysic. It is not by chance, she notes, that the nature-endorsing *Weltanschauung* eventually held by said theologies and further philosophical critiques is associated with biopower and opposition.[7] Both are rejections of the ills of modernity with its technocratic Prometheanism, anthro and andropocentricism, instrumental rationality derived from a

---

6 Don Cupitt, *Mysticism after Modernity* (Oxford: Blackwell, 1998), pp. 41–43.

7 Kate Soper, *What Is Nature?* (Oxford: Blackwell, 1995), pp. 119–49.

scientistic model assumption on the superiority of humans over nature.[8] This theological critique, she notes, is likewise carried in the philosophies of Derrida, Foucault and others in which there is a communicational and certainly ideological impasse between the natural and the unnatural, built and unbuilt, individual and species and so forth. Humanity as independent from nature, the goal of modernity, is thought to be complete when in fact, as the critics note, it is embedded into nature or more saliently its imposition on it. We cannot but help to read our political and power ontology into nature, but it is always ours and not eternal. Hence, we are of two solitudes when it comes to nature. One solitude is the modern assumption of nature as discourse independent of us, reduced to natural laws that clever but blind human beings have learnt to manipulate (nature skeptical) and the other solitude is nature as a signified signifier, constructed by power modalities and exploited therein, *but* which has some *metaphysical privilege* in it if we learn to parse the signs accordingly (nature endorsing). This is critical to grasp—the assumption of control or exploitation of nature, including human nature, by power structures presumes the presence of a tacit exploited metaphysical connection and to which one must return to stop the proverbial cultural rot. Thus, modern Western society has a double job when it comes to refounding interhumane connection—it must defeat the exploitation of nature by social powers that are parasitic to general human supranature connection and stop the exploitation of nature by science that denies that impulse in favor of Darwinian competition and chaos. More simply, the one is science as reduced and the other is a return of the spiritual native or in my case the consilience argument.

Clearly, for the disenchantment view, nature endorsing is more attractive as it places humanity squarely into the conversation of what is nature and presumably exactly how we relate to it. Its strength is as antidote to Ulrich Beck's claim that we have somehow moved into risk management and that we continue to consider ourselves independent when in fact we are dependent within on even on nature.[9] Over the past decades, many theologians and philosophers have dedicated much thought to this awareness that humans are indeed within nature, that to think we are above it is the root cause of much of our environmental future, admittedly bleak. It seems so persuasive an idea—that we are in the Disneyesque "Circle of Life"—that is accepted easily as an idea. Being natural is better than being above nature, and the problem of the last Christian centuries is our arrogance toward nature as its master. Of course, there are also more extreme nature-endorsing locations of pantheism

---

8 Soper, *What Is Nature?*, 1–6.

9 Ulrich Beck, trans. M Ritter, *Risk Society: Towards a New Modernity* (London: Sage Publications, 1986, 2005).

and nature religion, independent of attempts to recover within the Abrahamic traditions a "green God."[10] However, is this correct in its basic approach? Is the idea of Nature endorsing a better alternative to the master narrative of both classical theism and science?

Soper's usefulness is to underscore a simple point. Nature endorsing, for all of its promised inclusion, is as infected with the same self-dualism (the spiritual as better than the material) that "plagues" the science camp and as a result fail to understand exactly how humans are in nature and above nature because both camps explicitly place the human, and human exceptionality, as a priori any definition of nature.[11] There is no difference between the scientist who sees nature as something to act upon by understanding and manipulating causality and the environmental pantheist/theist who argues for a respect and consideration of the ecology of the world. Tracing the relationship between nature and the human, Soper argues that what happened was a cosmological shift in agents (God then human) but not in an *essential centrality* of the human person in that cosmological drama. Whether Descartes, Kant, Hegel, Heidegger or Derrida, *homo sapiens* is always a subject in the object (which is nature), what varies is whether this was modernity's *individuum* or postmodernity's collective[12] and then which sign is privileged to decode the symbol.[13] But the endpoint is the same, the "*nomos*" (culture) of human nature is confused with the "*physis*" of natural reality and humanity hovers over both as a cryptospeciesism. In both cases, humanity is simultaneously credited with far too much, and far too little.

Another way of understanding this is this: the emergence theorist is too far indebted to the scientist and that gives away too much as both start at the same place theoretically. It is in assuming that humans are in some centralized, and therefore privileged, account of reality. The dualism here is not the classic dualism of mind/body but rather a dualism that assumes some part of humanity—whether it is consciousness or morality or just our ability to see nature itself in science—has some ontological power that allows pristine and unadulterated privilege as access to reality and the Real. It is easily understood that the nature skeptic, using scientific reasoning, sees nature as a kind of Real that humans can access without issue. It is much more nuanced, however, to consider that the nature endorser does the same—that there is something in the human, that founds Nature itself, that allows a likewise interdependent but

10 For example, Sallie McFague, *The Body of God: A Theological Ecology* (Minneapolis, MN: Fortress, 1993).

11 Soper, *What Is Nature?*, 39.

12 Soper, *What Is Nature?*, 46.

13 Soper, *What Is Nature?*, 57.

correct reading of the signs of Nature. If both share this flaw, then can there be a way forward?

Where both ideas falter is at what Soper (borrowing from Ted Benton) calls a "deep structure" understanding of Nature. Deep structure is the idea that there are inherent laws of causation that come up from reality itself (as the scientific nature skeptical holds), and an understanding of that causation that comes from equally complex human activity—including its supervenient needs or collectivisms and actions, moral and otherwise. In other words, nature is causally independent (or ontologically simple) so that if the ozone layer has a hole, it has a hole because of X and Y and its impact will be A and B. Humans are involved as a part of nature in that causation as inherent in nature and its laws (e.g., we participate in it and are acted on by it according to Nature's laws that even determine us). A hole in the ozone layer means certain radiation comes through from the Sun, and this in turn, causes human skin damage amongst many other things. However, humans—perhaps by evolutionary accident or divine purpose—also impinge above nature and use nature for its own purposes and its "laws," which in turn emerges with a new kind of complexity dictated by the inherent causal laws of the world. Thus, geological nature may in fact create gases that attack ozone molecules as a by-product of fundamental geochemistry and so forth, but humans also make some flourocarbons that accelerate that "natural" process, and in turn, alter the environment and themselves in doing so in ways unexpected or predicted by the laws inherent in causation. Taking something like climate change, it can be both blindly nature as a part of Nature's natural processes, but also something that is anthropocentrically caused. Deep-structure understanding tells us the simple matter that we cannot alter the fundamental structures of causality in nature—humans are entirely natural—but also that humans can in fact alter *physis* with our *nomos* or culture. This is both in terms of real action and in terms of interpretation. We do things, think in certain ways and both are placed into the mix of how we are—really are—in the world. Causation is ontologically natural and epistemologically learnt—but they are both two different magisteria (*physis* and *nomos*) and there is but one singular real, that of Nature. This is, of course, the preoccupation of those thinkers who fix on methods and models in the relationship between theology and science.

The import of this, if I am correct, is quite blindingly simple. A deep-structure understanding for our purposes of neuroscience and theology must adopt a realism that acknowledges the deep-structure dilemma of speciesism and causation. Nature is not something outside of humans, we are nature and what happens in and of nature will indeed happen to us. This is the world of the scientist—to attempt to tell us what the causal laws are, and how those laws impact specifically humans and nature in general. This goes all the way to the ground—it is ontologically simple reductionism. However, we are also speciesist, in that uniquely humans create tribal collectives that have "emergent" properties (how we think of ourselves as more than nature) coming out of nature (including our nature), and that those collectives or tribes are motivated by signs read back into the symbol of Nature. This is the world of the ethicist, the clergy person, the parent and so forth. A simple contrast is illustrative. A weed does not think that overtaking your front lawn will destroy its own existence as it depletes the resources it needs to thrive. The weed's nature (*physis*) is to grow but it does not have the emergent supranature to deny growth (*nomos*) for the sake of future events (except that selected by *physis* as evolution). It is completely natural. Humans are not. We are both, at least in construction of awareness of the role our *nomos* has on our *physis*. The truth is that neither can be separated for human beingness—what we do—without violence to human uniqueness, and more particularly, to human volition and mores. To consider altering our downward causal effect, we must retain human uniqueness as an imagined space to unify will to action. More simply, we need to construct *fictions* that allow us to see ourselves downward, as coming to understand our place in Nature and not merely as natural alone.

The implications are clear. The first is a consideration of what kind(s) of metaphysic drives the moral bus. If we have deep-structure understanding, then they cannot easily follow into the model of the pure nature skeptic (and use its tools only) but certainly cannot be nature endorsing as this leads to a pantheism or panpsychism. We need to be, ahem, speciesist. But we cannot help but be speciesist—even the most ardent scientific reductionist such as Dawkins must exceed biology. Thus, we will *interpret reality* as we do, with signs and symbols—as a metaphysic—and some of those will come from theology or religion until we abandon them with a new *nomos* to explain our *physis* (This is our last chapter). But, for now, we must also consider whether such a metaphysic that *prima facie* denies the nature-skeptic model is coherent at all. If the metaphysic contradicts basic reality, then the question becomes what kind of *nomos* is inherent in the said failed consilience because such opposition will surely fail to allow *mores* in a realistic fashion. If we are entirely natural, then we do have to deal with this reality (say in legal theory which denies aberrant behavior as physiological nature). Finally, we must ask in the case of the religious-motivated consilience movement, whether its inherent commitment to a kind of metaphysical nonreality placing (human) nature as a demigod is in fact at all coherent for its community and traditions. The God of nature—even our own pure nature of soul—will not do better than the God of the bible in motivating just or moral agents if Soper's thesis of humans as part of nature is really, and as argued by Murphy, a form of pantheism/panpsychism as the human is a kind of world soul. This kind of fundamentalism will fail to capture the rationality of ethics as it endorses a sanctity and authority within nature that simply is not sociopolitically possible for the majority as it places few in the place of interpreting the nature of that common world soul.

The last consideration is more difficult to follow as it asks a more fundamental question raised by the disenchantment thesis. The implicit argument is that something—particularly something holy—has been lost in the modern world due to scientism and its ills. The solution for many religious communities is an antiscientism stance in which space is carved out for the holy in the gaps left by science and a queer understanding of science as a perverted antinatural discourse of sign. Again, while there may indeed be *some* truth to the claim that scientism can and do, for some, operate as a kind of religion and metaphysic, it is more to the point that the claim of reductive scientism is largely a straw man and derives from a misunderstanding of science as a discourse. Natural skepticism is not, by definition, the same as reduction to scientism with its lessons of social Darwinism, *Frankenstein* ethics and sterile unconnected worlds of *1984*. Not even Mr. Spock or Sheldon Cooper can remain islands of rational, discursive reason—if episodic television is at all truthful (wink).

Rejecting Reductionism

Many Christian (and Jewish) theologians on science are quick to run to two fundamental places to put science and its largely material and therefore atheistic understanding in its place. The "God of the gaps" theory is used to fill in tautologies found in science, and especially those between quantum and classical physics, and as I argued, consilience. The logic here is that claim that everything will be reduced to a simple equation is patently falsified by science itself and that emergence instead allows for the possibility of explanations beyond simple reductionist causality. Presumably, God is one of those explanations; and we return to the classic two competencies or magisterium of science and metaphysics with each respecting the other's sphere as one cannot explain everything. The second tack is a crypto-"God of the gaps" theorem used by more able theological minds and squarely aimed at evolution (e.g., Nancey Murphy) who doubly attack the "barren religion of Scientific High Priests" and their sin of reductionism in areas of human behavior and other macrosystems whose principles are clearly emergent—whether in terms of sociology, evolutionary psychology or biological systems including the microbial world. This is Soper's thesis above on the confusion of *physis* and *nomos*. Here the attack is that old dualist metaphysics, remnants of "Descartes error" to find a mathematical certainty has blinded science to C. S. Peirce's critique of Tychism wherein what science produces is not fallible inquiry but falsely immutable deterministic laws. The argument, then, argues that if science is fallible inquiry, and add modern spins on it as a kind of biopower discourse of sign and the like, then it must instead be modest and include possibilities of other kinds of discourse including discourse on an organizing immanence whose goal is ensouled humans. Like the first tack, it ends ultimately with a kind of two competencies theorem but does so by enhancing one kind of science against another kind of science (theology), which has a commitment inherent to a world soul, emergent *conatus*, and privileged humans. It is a subtle argument, allowing a possible space for God (or at least dialogue) as an explanation, and certainly bears the respectability of scientific method, logic and argument to its limits. Soper, although she does not directly engage with this camp, could identify the dominant trait of this group as belonging, as the scientist, to the nature-skeptical position. The nature-skeptical camp, you recall, treats nature as an object independent of the human subject, and one which, given enough information, the human subject can manipulate to said advantage so that it must also recognize limits as to what science can say about Nature. This is a powerful postfoundationalist thought, the need for science to interact with other human disciplines, including theology.

The problem with both these camps when they come to multivalent phenomena is simple—the argument that science is both reductionist and human agenda driven neatly combines into a conversation that essentially argues either conspiracy (say in vaccines), misread of data or blatant political manipulation.[14] This misread of science and its confusion between ontological and epistemological reductionism allows a political agenda that exploits culture wars and uses pseudo-science as a wedge issue to move more fundamentally religious questions within the secularization thesis. It is not about the limits of the discussion, but rather the removal at the beginning of claims that the nature skeptic might make as the necessity for a new consideration of *nomos* based on *physis*. It is, by definition, always an incomplete activity as the deep-structure understanding of Nature is to include humans in that understanding as a messy contingency using the tools of science to pare away false ontologies. Note, I did not write using metaphysics to pare away false ontologies.

However, as even Peirce (himself a committed panpsychist) noted, science is really about probabilities and his "reduction thesis" holds true for how science manifests in terms of method, and more particularly in terms of application. Peirce's insistence that triadic relations (and more importantly Quine's disproof of it in favor of dyadics) as almost always the better option of explanation is not the same as disproving ontological reductionism but rather a cautionary take against confusing epistemological reductionism and its relata

---

14 See, for example, Michael Mann, *The Hockey Stick and the Climate Wars* (New York: Columbia University Press, 2012).

with fundamental reality. Ontological reductionism is the simple claim that the most fundamental physical level is ultimately "Real," and that anything else considered real must be mapped onto or determined by that fundamental ontology.[15] Epistemological reductionism is the confusion that theories or schema about the macroscopic features of the world can be reduced to or mapped directly to the Real.

Ontological reductionism operates in an entirely Peircian and Quinian manner by using elimination (not X but Y), identity (X is really Y on understanding, or a subset of Y), mereological supervenience (Z is the sum of X and Y but cannot contradict properties of X and Y) and finally nomological supervenience (Z is bounded by a set of properties of X and Y and includes a series of possible states of Z within). In this case, it can be considered that for the most part, even with the complexity of systems such as the brain, the goal of the scientist is to identify properties at the most fundamental level to determine the structural relata therein. It is open to investigation, public and can even be overturned albeit reluctantly as humans cannot help but read their *nomos* into their *physis*. It undergirds the most historic classic basic belief of science that is indeed quasi-religious—that of a *teleos* and *ontos* in nature. Of course, whether we can truly get there is an entirely other matter, and the previous chapter's summary of CQFT and quantum mechanics is the poster candidate for this quest, anxiety and primacy of vision. It is the goal of science in its quest for a GUT. But it is also open to both influence and operates by Popperian falsification as a kind of Inference to Best Explanation in its simplest form. It does not, however, presume domains to be separated by their competency (as cold fact versus compassionate morals) but does assume a foundational reality that includes the human *nomos* as a feature of *physis*. Ontological reductionism believes there is an order into which humans, including as quantum reality and conscious beings, are placed and that this order is sufficient to understand FAPP. *It is education to reality.*

This being confused with epistemological reductionism is where problems arise but need not if properly understood. Epistemological reductionism is our attempt to organize the relata within ontological reductive statements and operates by replacement (Theorem T1 is replaced by Theorem T2), theoretical derivation (really Peirce's triadics) in which Theorem T1.2 is derived from Theorem T1 creating a new relata so that what is said about T1 is also in T1.2 but T1.2 explains a new set of relationship specific to T1.2. These are not homogeneous but heterogeneous in that T1.2 is a specific set of relata

15 Michael Silberstein, "Reduction, Emergence and Explanation," *The Blackwell Guide to the Philosophy of Science*, ed. Peter Machamer and Michael Silberstein (Oxford: Blackwell, 2002), pp. 80–108.

not discernable alone by T1 (e.g., temperature). Epistemologically reduced statements have domains where they work and are not ontologically true. The classical difference between classical physics and quantum mechanics is an example of epistemological reduction—classical physics work for our universe's energy domain, and quantum mechanics is the attempt to see the ontology behind that, extending beyond our energy domain. The latter is true ontology, the former epistemological reduction. The belief behind science is to push to the ontological and therefore can be either semantic or pragmatic epistemological models. Semantic epistemic models consider scientific theories as possible mathematical models chosen for their utility for the relata—hence, Euclidean geometry works for Newtonian physics, whereas advanced mathematical such as algebra/calculus/field theory formula work for quantum physics because they describe different expressions of the ontologically simple. Pragmatic reductionism is a dump all place, essentially arguing that scientific theories are merely linguistic entities and work for their specific locations. Hence, I would use Newton to shoot a rocket, and quantum theory and special relativity in understanding Big Bang cosmology. Neither, on the outset, would involve questions on the morality of the rocket or the Big Bang. These, of course, come naturally enough as humans ask just such questions—what does it mean?

What is salient for our purposes is that epistemological reductionism, particularly of the semantic and pragmatic kinds, is the misunderstanding of intertheoretic synchronisms posed by ontological and particularly mereological and nomological supervenience chased by the specter of Peirce's discredited dualism. In other words, theorems may not be simply mapped one to one onto reality and that it is here that interpretation occurs and can be the motivating factor. But, and equally important, these does not by any logical consideration, as argued both by Peirce and Quine, necessitate that say emergence is true just because we cannot explain the complex interactions at the higher end of the causality or the theories interactions. In other words, our semantic limitations are epistemic symbols of the deeper ontology that we share reality in, and these limitations are not, by definition, reality itself. They are our "culture," but they are also not merely our "culture." To reduce all to the same node—reductionism—is to misunderstand how theorems work moving from epistemologically reduced models that are semantically bound as expressions to ontological simple reduction. Core theory, we recall, is FAPP!

The point of this then for neuroscience is simple. When scientists make claims about the complexity of brain and identity and its causes and impact, that they are speaking semantic and pragmatic theorems is true, but this does not, by definition, mean that they are also not speaking of the Real in its most fundamental causal relata. Further, I would expect then that theorem or models are prone to political human manipulation, but I would also add that, done correctly, science as a specific kind of discourse is driven by its quest for ontological reductionism, which can replace the relata of false epistemic articulations or derive new relata from incomplete or local articulations. What we really have, then, is not a problem within the data but a problem of interpretation, which is also a problem of will and culture in which there are hidden relata and what is privileged in those systems. To argue, as in the emergence crowd, that science is unreliable *de facto* is a misunderstanding, perhaps even by both sides, of what is the method of science and ontological reductionism—which is its bedrock—is not the bogeyman it seems.

Making Decisions

If the problem is a hermeneutic one, and not a data or evidence one, then what can we do? I suspect that much of the issue in play is about the moral features, and the idea that we are free to make our moral choices as a modern phenomenon. In this case, the problem is that theology is held hostage by its commitments to modernity rather than its theologic. Contemporary moral psychology is a reflection of a general conversation held by the Western world between cold rationality and nativist intuitions with reason/logic aligning against emotion/intuition followed by the critique that we are socialized through education.[16] The modern mode is that education in right reasoning allowed for unbiased morality (metaphysics and morality), followed quickly by a response that empathy, placement of the other in ourselves, was the proper place of education and that connectiveness in its various forms was essential and this was in turn the goal of education in morals. Of course, religion floats nicely in the tide pools, with moral *habitus* being either ontologically present in nature, human nature, or some divine intrusion into creation of which institution inculcates as a by-product of its mandate. What did happen, however, was that cold dispassionate reasoning fell off as few used moral calculators and fewer still saw in the Habermasian public sphere the free exchange of moral and rational ideas. It seems we make moral decisions based on education and bias. Reason is the rider on an elephant of moral preconceptions, not the elephant itself.

Thus, as Christopher Haidt notes, when psychologists began to explore morality in the 1980s onwards, investigators found religion (and new religions such as modernity itself) to work as a Durkheimian social convention for moral decisions (e.g., Kohlberg, Shweder and Damasio) in which it was less

---

16 Jonathan Haidt, *The Righteous Mind: Why Good People Are Divided by Politics and Religion* (New York: Vintage, 2013), pp. 3–72.

rationality in play, but more intuitive or biased learnt orientation that informed moral reasoning. Moral reasoning, particularly in areas of harm/injustice, taboo and disgust/revulsion, work to reinforce basic tenets of allowing a society to function as *that* society without necessary eternal or ontological reasons—these come afterwards. What research has indicated is simple: moral reasoning comes less as a persuasive technique and more as a reinforcement of already held convictions, and that moral reasoning is culture specific in the definitions of harm/injustice and taboo/revulsion.[17] As Haidt suggests, "religion is a team (moral) sport." The source of this, Haidt suggests, is that we are evo-social and that our evolution as different primates is also an evolution of religion as a handmaiden or moral exoskeleton to articulate and guide our hidden elephants.[18] This also means that such evolution is not a supranatural force or appendage we share with Nature itself, it is a *nomos* in that *physis*.

The implication of this is clear. First, whatever arguments serve our learnt elephants are likely to be more persuasive to us as courses of actions but fall deaf on the ears of other elephants, even if our rider is sharply dressed and morally persuasive. Conversion is nearly impossible. We riders stick to our tribes with deep devotion, regardless of whether the other tribe looks attractive. Second, and more important for the purposes of this paper, the elephant is more important to understand than the rider, and yet we focus on the rider in the debate. This is the case in terms of neuroscience and theology—we focus on the rider (what the science may tell and why religion bucks it) instead of asking why that theology itself may be infected in the first place. If we are truly interested in being educated to reality, then we need to look to the elephant that is reality as we have come to know it, not as we have inherited—even if that was a largely good inheritance that served well. We need to learn to pay attention to the elephant, and this elephant is "us" as creatures in social networks and as "brains" as well. *This is education to reality.*

Haidt gives us an important help in identifying some of the elephants in gross terms and does so without paying much attention to religion per se (although it is clear three are largely found in religious moral discussion and three in secular discussion). He identifies care, fairness and liberty—really the idea of the individual before society as the keystone to understand morality as empathy or the golden rule—and their religious counters of loyalty, authority and sanctity, which all serve collective goods. Consider this a dialogue between Jeremy Bentham and his Panoptican society and the natural law or rights of any society given justification by divine fiat or some other deontology. Haidt's

17 Haidt, *The Righteous Mind*, 189–274.

18 Haidt, *The Righteous Mind*, 221–76.

care, fairness and liberty are really nothing more than utilitarianism's principle of utility, whereas loyalty, authority and sanctity serve to reinforce altruisms to curb unnecessary usurpation by utility.[19] Haidt, speaking in his own American context, identified the political left with the attributes of reasoned utility and the political right with sanctity and thus he surmises that in collective decisions, the elephant predicts the rider's choices. Thomas Frank makes a similar kind of argument in his book *What's the Matter with Kansas*. Frank argues that recent political debate in the United States is a response against the rapid deconstruction of perceived sanctity of the American mythos by a liberal elite around some moral areas such as abortion and homosexuality, and that as such these issues serve to become the elephant on other choices in nonrelated spheres of morality such as economic regulation. In other words, because many feel that the values of American family are undermined in this sexual area, their moral reasoning on other issues such as banking, government and so forth are driven, even against their own utility, by those wedge issues because of the emotive value of sanctity, loyalty and authority, whereas liberal secular voters approach those issues from the perspective of care, fairness and liberty.[20] Recent elections in the United States reinforce this even more sharply. The 2016 YouTube videos of the Slovenian philosopher Slavoj Žižek on the election of now ex-President Trump argues just such a point.

What Frank and Haidt hint is that the two solitudes may have two approaches to decisions about what to do in response to neuroscience as claimed and more perniciously, that the responses may in fact have very little to do with the data of neuroscience (or any science) at all. As well, if Soper is correct, then the understanding of nature—either skeptical or endorsing—is also the rational versus emotive poles once more and as such will vary in response. What is needful is an approach that allows both emotion and reason, and nature to be understood correctly. This I propose is a nature-realist position. The nature realist is the end of the past centuries of conflict, definition and concord between theology and science. It is, I maintain, our modern minds finally leaving our Stone Age souls. *It is education to reality.*

A Philosophy of Nature Realism, the Role of Religion and the Image of God

The nature-realist position has the following attributes: (1) It tries to understand nature as something into which humans are firmly entrenched

---

19 Haidt, *The Righteous Mind*, 274–315.
20 Thomas Frank, *What's the Matter with Kansas?* (New York: Metropolis, 2004).

as natural beings linked casually to the whole and (2) understands humans as having a unique position within nature that allows human beings to consider the impact of their actions within the whole. In philosophical terms, human beings, when it comes to nature, are both object and subjects and that it is entirely essential for our species to exist in both modes as subjects within nature and as objects within nature. I have argued that the issue in discourse around neuroscience is that the false dichotomy of nature-skeptical (nature as object) and nature-endorsing (nature as subject) or reason versus sanctity makes volition useful only to those already within that framework. Treating nature as merely something external or merely something of value to us reduces the chances of persuasion and increases the reality of culture wars by a *legerdemain* of values transposed politically so that the discussion of nature, and specifically science, is really an attack on religion, science, secularization or hermeneutics or another elephant.

Soper introduces the idea of nature realism in her work but I want to extend it. She argues that the metaphysical or nature-endorsing thinker who sacralizes nature (and by implication humanity) makes it too difficult for most to worship at the altar of nature as their metaphysic or sanctity lay elsewhere, and particularly the god (or false gods) of modernity for both the scientific and religious inclined. The nature endorsers see in nature a reflection of their own face, and that face wears the individualistic freedom of modernity. The pure realist, the nature skeptic, gives the information, places humanity squarely in nature, but is exploited by its own multiplicity and beliefs in technological solutions or mere stoicism. The nature skeptic sees themselves, at least in deep structure, as above nature with godlike vision. What she argues, and where I agree, is for a plurality in modes of action and that it is both the rational and the aesthetic that needs to be engaged for action. It matters not whether one is active because of utility in self-preservation or aesthetic in connectedness to the natural, just that both are at the table. The problem with her position, however, remains the two solitudes with little interaction albeit some hope that both can row in the same direction albeit for different reasons. The real issue, however, I find with her position is that there is no substantial reason for changing sides of the bed for the noncommitted or resistant. The issues within nature skepticism and nature endorsing remain the same—both serving idols that fail to persuade. I go further than she in that I argue the nature-realist position must also hold to a substantial critique of hidden values.

Those hidden values, I maintain, are initially more important to be discussed than the scientific data as these are the locations of the culture wars into which science versus or in theology is placed. Here in the "Christian West" is where philosophers and theologians can contribute to the work. The work is simple: (1) to do the hard work of exposing idols that haunt the conversation, especially antiscience as a response to secularization as well as the equally dangerous idol of re-enchantment; (2) to help nature skeptics understand that human exceptionalism is not going to shift from the center of most discussions within science and that this needs to be part of any engagement because the idea of sanctity is primary in moral reasoning. In the Christian West, the sanctity of the human person as a unique species is hardwired (even in nature religions) and to present data without appealing to this aspect of human life is to speak into the void; and (3) to help nature endorsers equally to understand that nature is something humans are not necessarily placed into with special stewardship or communion but that humans are in nature and will indeed alter it as any species in the general evolutionary process. This desacralizing of nature allows for reasoned alteration of behavior, justifying steps that may not restore what was but certainly can alter the worst of impact. But there will be impact. If this is the role of religious or philosophical thinking to open the doors for Soper's plurality with the emphasis that without persuasion there can be no volition, then the question is how.

There is, as a very brief aside, a place that seemingly has worked in theology, and in particular Christian theology, within the neoorthodoxy associated with the early twentieth-century theologians' crisis theology, and their progeny. But to spend time here would be to detract from a very dense argument already engaged. Suffice it to say, that here there is a tool theologically for the devolution of modernity's centrality of the human person as the subject of theology and the founding of a discussion on transcendence. However, whether the split can go further than Karl Barth and Emil Brunner's split of the 1930s, with I think Brunner carrying the day writ large now, remains an open question.

The process of identifying our gods is really the process of maturation. It is distinguishing our fictions from our delusions and then constructing a morality of sentiment accordingly. The final chapters explore more fully this idea: How does one become a nature realist and which roles can theology and science (as above) play in this conversation? More succinctly, *how does one educate to reality and distinguish between fictions and illusions if we are forever enmeshed in our own perceptions of the natural?*

Chapter 5

RETURNING THE SOUL: FICTIONS OR ILLUSIONS?

In 1927, at an axial time of European history, Sigmund Freud published his *The Future of an Illusion* as a completion on his work on the origins of religion, started in 1913 with *Totem and Taboo*. In the work, Freud argues that religion is a necessary fiction whose illusionary power allowed humankind to become civilized while balancing primal/animal instinct of the self with the needs of living in a collective. In the work, Freud claims to bring psychoanalysis, psycho-development, and other disciplines such as archeology and anthropology to understand the origins and prevalence of religion especially that associated with the Judeo-Christian tradition. The themes in the work would persist throughout his later life, including his 1939 *Moses and Monotheism*. While it may be accurate, as scholars repeatedly argue, that Freud's work is dismissive of uniqueness of monotheism (and perhaps gives too much credit to the simple reduction of psychosexual development as the reason for it), it is also true that his claim that religion is a fiction whose illusion allows human society to make ontological claims on the nature of moral and natural reality rings true with most contemporary sociological theories on religion.[1] Religion, particularly in light of the many historical and present religious candidates of human experience, at its root has a sociofunctional role and that this sociofunctional role can be the reason it is accepted, dismissed or vulnerable to usurpation by another religious identity over time. In short, most sociological analyses of religion find that the wide variety of ontotheological claims—whether Islam or Zoroastrianism—are best understood by the role of social stability that the worldviews grant. This position, on the surface, makes nothing of the underlying "truth" of the divine reality/ities but only that they "work" in granting social utility, both communal and personal. This means, using Freud's language, that they are "fictions" granted power by their utility and as such make no claim on ontology

1 See, on this, Harvie Ferguson, *The Lure of Dreams: Freud and the Construction of Modernity* (New York: Routledge, 1996), especially pp. 194–225; Anthony Elliot, *Social Theory and Psychoanalysis in Transition*, (Oxford: Blackwell, 1992), especially pp. 235–73.

itself. In words, they are made up by humans as they understand themselves, not necessarily a true investigation or understanding of the Real. They are "illusions" that grant social stability (and by extension, social abuse), given ontological power by their usefulness and therefore entirely human—perhaps local but certainly not divine in origin. Freud, it seems, was prescient in this way and outside of deeply conservative theological or religious circles, one would be hard pressed to find a defense of religion and its claims on reality that ignores this human coefficient and practical evaluation. Whether the role of women or gender, the spectrum of human sexuality or the interdialogue of seemingly incompatible religious claims, most of the world's living religions are evaluating or altering traditional dogma in new human experience and understanding of reality, even if only in rejection of those thoughts.

Culture, Civilization and Gods

Freud's account of human personhood is controversial. For many, he grants too much to the hidden subconsciousness self, is preoccupied with sexuality and instinct and is not at all scientific in the way that the general psychological *sciences* have evolved.[2] All of this is true. This is not a defense of Freud's work as a work of psychology per se, but rather a defense of his assertion that religion historically functioned sociologically for the most part, and that this sociological function had a personal yield in terms of allowing people to reconcile two realities—the social and the natural person. By the social, what is meant is that cultures come together and establish social ideals that regulate all aspects from birth to death and include economics, politics and morality therein. By the natural person (instinct), what is meant is that beneath this veneer of social ideals, the human person is driven by our natural desires, everything from basic survival to our desire to reproduce—and that perhaps, we are just a particularly self-conscious (and deluding) animal in this regard. We live in the collective ideal as individual cases of limited, biological self-understanding and expression. Freud's terms for these two realities have a loaded meaning today and can be associated with a kind of Western imperialism and Orientalism. Nonetheless, they can be helpful if parsed carefully. Freud's term for the social is civilization, and his term for the natural person is culture.

Civilization, for Freud, is the collective realization that human society needs to establish some ground rules for itself, or inherent logic to secure peaceable life together. This is its ideal ontology—its gods or God. This may be a kinder

2 See, Tracey Hess, *Consciousness and Quantum Theory: Freud's Death Instinct and the Phenomenology of Corbin's mundus imaginalis Manifested as a Complex Fractal in Epilepsy*, unpublished PhD thesis. Pacifica Graduate Institute, 2004.

society, but it need not be. It can be powerful and coercive, exclusive and imperial and terrible to the other. It may justify but a few, or it may be benign. But it is arbitrary or local and certainly not a part of some deep universally organizing ontological moral order. In fact, Freud explicitly notes that it is more likely to be built on "coercion and renunciation of instinct"[3] than not. Freud, after all, lived through the Great War. Culture, on the other hand, is that instinct that we need to learn to renounce, and while this seems oxymoronic to a contemporary reader (is not culture just not another word for civilization?). Culture is the way by which an individual learns to bend, modify, mollify or use personal instinct or interest to live in civilization. Culture is both our natural selves and that natural self learnt as reconciled to an external logic of a greater social, historical and collective reality of civilization.

Culture is the decision(s)/identity that any one person (or small group) makes as they balance their own desires with that of fitting into the much wider communal life. Culture, for Freud, is the decision to "play along" with the social mores, the social norms and the social expectations while navigating one's internal psychological and instinctive life—including our biology and desires as well as social imaginations. Of course, this bears similarity to Freud's ego, superego and id which, at its most simple, is the self (ego) navigating self-expectation as mirrored by the external others (superego) and one's inner life and desires (id). The imbalance of these three aspects of one's personality is the source of delusion and psychological pathology. Civilization here is the superego, culture is the ego as the id's public face and navigating between those two fictions is what our instinctive natural self does.[4] Freud is at pains to argue in the essay that navigating between civilization and culture is the source of "frustration" for most—our need to live collectively together and our internal lives that may chafe at the norms established by the collective and our manifestation of those norms behaviorally as culture. The most successful cultures, he argues, accrue the greatest "mental wealth," by which Freud means the closest concord between civilization and culture so that the greatest number of persons can agree to the culture as civilization. And, not surprisingly, the greatest item in this inventory is not economics or art but rather religion:

> No mention has yet been made of what is perhaps the most important psychical inventory of a culture. This consists in its religious ideas in the widest sense—in other words [...] in its illusions.[5]

3 Freud, *Future of an Illusion*, 7–12.
4 Freud, *Future of an Illusion*, 18, explicitly.
5 Freud, *Future of an Illusion*, 24.

Religion is mental wealth. It allows the greatest participation while yielding the most benefit overall as it addresses the basic id, the self-constructed ego against the civilized superego. In short, it works, has worked, and should continue to work. Or should it? Has something changed?

Freud is hostile to religion, not because of what it "historically" did as granting social cohesion and stability, but rather because it is no longer necessary because humanity understands its relationship of culture and civilization differently at present. Religion was necessary in a prescientific world not only as it restricted basic instincts, primarily sexual and violent (but these, for him, sublimated), but more importantly because it gave the gift of "humanization of nature." This humanization of nature took the greater natural world, including our own mortality, which our finite, material bodies suffered from and with, and gave nature (including our own) purpose and explanation. This projecting purpose, and kinship and even care, to supranatural agents provided relief for us by making the agents similar to us—they could be "adjured, appeased, bribed and influenced" and robbed of power over us because these agents were *our culture*.[6] They were us, and we, in turn, like them. These collective illusions were borne out of fear of nature, made nature supranatural, and in doing so of our "culture." This is the world of Christ Church Cathedral in Oxford with its demons, angels and humans that opened Chapter 1. Again, Freud finds this parallel to his emerging theory of psycho-development, and this is the infantile stage of development. Freud borrows these ideas from the emerging discipline of neuroscience learnt from Hermann von Helmholtz and Theodor Meynert's "German brain model" in which both rejected dualism metaphysics (shared by Schopenhauer) and instead believed that the world as represented and as "is" are the same but are also representations. The brain was a machine that experienced the world, contra heady Kantian metaphysics. Importantly, for Meynert, this representation could be interfered with, due to either development or abnormality. But it was the world as experienced, not as idealized. Eventually, Freud would encounter this idea in Meynert's association of the brain and its growth as the "mechanism of civilization" through which brain's anatomical development (and damage or lack) allows representation of the world—including mores—in the mind to be a process of checking instinct and social. In other words, we are brain "all the way down" again. There is only representation—we are consummate fiction makers.[7] Thus, the brain's state conditioned how culture understood itself highlighted against the brief of civilization as an ideal.

---

6 Freud, *Future of an Illusion*, 29.

7 See, on this history, S. P. Fullinwider, *Towards a Rogue Reality: Kant to Freud, and Beyond* (New York: Peter Lang, 1998).

But if the supranatural agents were just projections of our culture, then how did they breach the ills within human desire that birthed them in the first instance? For Freud, infantile development of religion needed another stage of development that is the separation of the material and the Divine itself, as then the Divine can override or control that which created it. This, he proposes, is the gift of ethical monotheism, wherein the Divine stands as higher order within the disorder of culture. It is civilization being read back into individual culture, so that it can control the instincts within culture that may challenge the civilized self. It manifests as a series of precepts, and concepts, of which, for Freud, the most important is law and soul. They are related in that the divine law is the source material to which souls are aligned, and therefore above "human society, and extended over nature and the universe."[8] Thus, the development of religion is to introduce two parts of the human person, one of which has its source in the eternal independent of the source of discord within the human psyche that is our material basic desires. This internal higher purpose, soul, finds in the laws of revelation its own meaning, and stability is granted as it is not the gods made in our image, but our image made in the image of God:

> Life in this world serves a higher purpose; true, it is not easy to guess what that purpose is, but it certainly signifies a perfecting of man's nature. It is probably the spiritual part of man, the soul, which in the course of time has so slowly and unwillingly detached itself from the body that is the object of this elevation and exaltation [...] death itself is not an annihilation, is not a return to inorganic lifelessness, but the beginning of a new existence [...] to something higher.[9]

Freud has, then, a mechanism for the development of religion from polytheism to monotheism (and even to nationalism or if Marx, modernity and capitalism) that parallels his understanding of general development, transference of desire from mother to father, and his understanding of how pathology can occur in terms of the imbalance of id, ego and superego in the first two aspects of development. Religion, like all psychological reality for him, is a fiction—a constant reimagining of the self as one develops over biological and chronological time, as one balances collective identity with personal identity, and as one balances culture with desire. It has no ontology, no reality outside of the self and the collective as interpreted by the self. It is not the Real. There is no Real. There is only negotiation of fictions—representations conditioned

---

8 Freud, *Future of an Illusion*, 32.

9 Freud, *Future of an Illusion*, 32–33.

by our material existence (brains) in the material existence of other persons (civilization as manifest in its laws, social structures, etc.). Everything is, then, a fiction ultimately conditioned by what we would come to call biology or our nature. This would include everything from our biological limits to our evolutionary history. As a fiction, it is an illusion because it is not accessible because we are hopelessly mired in our own representations—a kind of mental Heisenberg Uncertainty Principle. But rather than reduce to nihilism, this truth can be a helpful one or a negative one as nihilism is absurd. Nihilism, at least for this writer, is the final solipsism and leads to suicide. That is too much of a price to pay for harsh materialism (more in the next chapter). So, the question becomes, are all fictions the same? When it is a negative illusion, it is a delusion. Later thinkers such as Lacan would parse this even more closely in terms of language, but the general idea that we have no access to the Real, but only representations of it remains consistent in most philosophy of the nineteenth and twentieth centuries.[10]

When Religion Becomes a Delusion?

Without doubt, Freud is a harsh critic of religion. Most religion tends toward delusion for him (as inappropriate knowledge of the self as an expression of a misunderstanding of the balance (culture) between the id (natural human or instinct) and superego (civilization)). Religion is juvenile or childish groupthink; however, given that most live in a state of immaturity, it was a best kind of groupthink. Nonetheless, the dawn of science has challenged this juvenile worldview. The reasons for his assertion are important—namely, that religion fails to explain the self as science has come to know the human person.

Freud's counter to the challenge that religion may be benign, socially useful, and even leading to peaceableness does not come from the extremities of religious delusion—fundamentalism, hypocrites and religious bigots—or from those necessarily suffering from being excluded because of failure to accommodate instinct to civilization—homosexuals, divorced person, mentally other and so on. Although Freud would likely be able to cite examples of each of the above as delusions enforced by, and perpetrated on, many persons in his care. Instead, the reason that religion is delusional as an illusion is that it fails the most basic test of being challenged as to whether it is a good fit for its job. In short, religion is unverifiable as a reality that it claims (and here is

10 The candidates here are multifold from Wittgenstein to Lacan and general postmodernism (see next chapter). For Freud and Lacan, see, Richard Boothby, *Freud as Philosopher* (New York: Routledge, 2001).

the ghost of logical positivism, enter stage left). Freud argues that its historical role as (1) already being proven useful by our ancestors, (2) proven by a theologic given by the conditions of that ancestry and (3) being judged as heretical or worse if one challenges the normalcy of the claims and logic means that religion could never be challenged or authenticated by the present needs of its possible adherents.[11] And, in fact, both historically, and experientially true now, those who do challenge such are frequently punished, excised or shunned as dangerous. For Freud, religion in the modern world was a mere shade of its previous explanatory powers. The world of science tested its claims, offered better considerations on culture and could offer a better civilization for all, including those marginalized by the status quo. Religion, then, is left with two possibilities. One is the *credo quia absurdum* mentality by which a believer holds faith regardless and the other is the "as if" argument by which the believer turns the tables on the skeptic and challenges the logic of the completeness of the skeptic's proposition or claims to the Real.[12] If we agree that proof is the bedrock of all statements of reality, then everything is challenged as if. Sound familiar?

To answer these two positions, Freud employs two tactics. To the believer who believes despite evidence, this is *delusion*. It is not an illusion that "works," it is a delusion that cripples as it creates more psychological privation than it heals, especially in a civilization that is not monolithic in its hegemony as the old Christian worlds, be they Protestant or Catholic. It may be a corporate delusion—and serve many, vast many, very well, but it is also punitive, restrictive and has no real "truth value" over a long period culturally and most certainly for many individuals in the short lifetime we are given: "ignorance is ignorance." Freud stands at the beginning of an axial moment of Western civilization, its assumptions are being tested as the world veers into the postmodern moments and the challenges to androcentrism, colonialism, racism begin to foment. Those assumptions, echoing now centuries of lament that interdenominational Christian polities created havoc for those living in them, are tested in the last world war, and the coming of totalitarianism that would haunt the West. For the "as if" philosopher, Freud employs a subtler and useful argument to answer his critics—he is indeed pursuing an illusion too. But the character of this illusion is that which follows Freud's understanding of the id/ego/superego in which the goal of psychotherapy is not the cessation of desires, of the warfare and questionings, but rather an *understanding* of them, which allows for balance rather than neurosis or delusion. For Freud, we are inveterate fiction makers, and what we need to do

11 Freud, *Future of an Illusion*, 44–51
12 Freud, *Future of an Illusion*, 49–51.

is to define that helplessness with a commitment to self-understanding, which he calls "education to reality."

Education to Reality?

For Freud, an education to reality shares the highest aspiration of religion—love of humankind and the decrease of suffering—and is, perhaps, still in its initial stages for Western culture but it has one attribute that religion does not—commitment to interest *in the world*, not the illusion *of the world*. This is an importance distinction. Interest in the world means understanding of the human person as natural, not as supranatural and as part of the world rather than the *teleos* or purpose of the world. In more simple terms, we are nature realists—not nature endorsing or even nature skeptics. Admittedly, he notes that this is not a very almighty deity and not much able to deliver what the old gods/God could at first as the servant of this deity—science—is still young and under development.[13] But, centrally, science as the handmaiden of educator to reality has four claims that religion does not.[14]

First, it is a human activity devoted to human understanding, and the only real tool humans can use to explore reality. This seems trite until one thinks that what Freud is gesturing is that very claim of vulnerability science has as a model of reality is because it is truly the only uniquely human activity of human minds to explore the world. It is, by definition, to be replaced, examined and scrutinized because of the investigator's fallibility, and has, unlike religion, no privileged place to sense, understand or receive higher truth. Science is our mental apparatus, just as religion once was but then religion, for Freud, made our minds as souls somehow out of the world—supranatural as a disguise for power. We are not fit for heaven, but this world through evolution. This is not a claim of the imperious nature of science, but rather that it is a human activity, using the human mind that is entirely natural in representation in the Meynert or neuroscience sense.

Second, science is part of the world it investigates and therefore can look backwards and forwards as it investigates its wider subject matter. To understand the world, ultimately, is to understand ourselves. To educate reality is to understand humans as finite animals who engage in illusion, fiction and sometimes delusion. We are entirely natural, and the world we explore is natural. There is nothing, including our own selves, outside of this. This is the whole game as it pertains to us. I use the word "game" intentionally to draw

---

13 Freud, *Future of an Illusion*, 57–58 and 68–69.

14 Freud, *Future of an Illusion*, 95–98.

attention to the claims of Wittgenstein on language and reality. Science is a game in this sense; it is our *nomos* and our *physis* as we label it.

Third, and following from the first two, science therefore has a modest task and limits. Its task is the world of our senses and understanding, directed back to our internal life (if necessary) as conscious, and its limit is to assume that our organization is that of the same world we investigate. There is no other heavenly realm in which there is some other essential Being whose theologic is different or undergirds reality, and to whom we are oriented or responsible. Science concerns us with reality, including our intramental reality as a manifestation of the reality around us, discovered by paying attention to it. This even, presumably, means questioning our own values, limits and histories.

Finally, and making science "no illusion" is that the ultimate findings of science—the destination rather than the process—means that humans can finally eliminate the abstractions created by our infantile worries of uniqueness, meaning and place in the universe. Freud's confidence in science as finally defeating the delusional neuroses of religion, economics, politics and so forth rests on the fact that initially science will be the psychotherapist for Western civilization and allow it to find a concord between culture and civilization, or the id–ego and superego. It will help us to find our place in the cosmos, to find a way to speak of our mental life—consciousness—as animals with bodies and desires, as collectives who wish to find peace, internal and external, and finally as a part of the very system it is both living off and destroying. This, of course, may sound odd but it flows nicely out of his own body of work in which he thinks the unconscious mind and the conscious mind are really two sides of intuition and abstraction, and that the goal of mental health (and by extension, social) is the resolution of the entropy (literally the death wish) of the two systems so that there is trust in the mind as onto-receptive as well as ontogenetic. Science, like psychotherapy, helps by eliminating the false realities of both the unconscious and conscious mind with a skeptical method.[15] Harvie Ferguson neatly sums Freud's understanding of science:

> For Freud "theory" is not "ego" of science; it does not gather together and arrange into a systematic unity; it is, rather, the "dream" of science, a mode of representation which, unrestricted by the exigencies of life, is sensitive to the multiplicity of meaningful perspectives immanent in experience.[16]

---

15 See Petr Bob, *The Brain and Conscious Unity: Freud's Omega* (New York: Springer, 2005), pp. 84–91.

16 Harvie, *Lure of Dreams*, 156.

Does this mean then that science can replace religion? No, for Freud, science cannot replace religion for humans are inveterate and necessary fiction makers, whose illusions are very part of navigating the world they live as individuals, civilizations, and species itself. Science cannot become the new religion—this would also mean it is prone to the same errors that it wished to replace wherein the abstract illusions (borne out of fear and need for control) created pull humans out of reality, including inner reality, and into some rarefied place such as nature endorsing pantheism or nature skeptical scientism. This will lead to delusion, and delusion to neurosis—perhaps the kind of eugenics or other superhuman genetics. This is the criticism of John Grey on positivistic scientism and its claims of totalitarian knowledge. Instead, science is a way to shape the fiction, to question it based on what humans are observing in the reality that they participate in and with, and to question what task and limit that cluster of illusions may demand as education to reality. After all, isn't neurosis famously the inability to hold ambiguity?

The Soul and Science

Given the claims of the fictions that science has provided in terms of neuroscience and quantum mechanics (not to say of evolutionary biology) in which humans are "enmeshed" can we speak adequately of a soul any longer? Can we hold with the ideas of a largely Judeo-Christian influenced soul that seeks communion with a Divine Being who ordained creation for such creatures, gave positive revelation to secure this and promised eternal life for such purposes? Can we think of ourselves as material plus? Is our universe merely a place local with hidden worlds within in which deity can secretly manifest in increasingly moments to a culmination of the future perfection? Will this temporal body be rejoined to a soul that persists in that future moment? The "education to reality" of both neuroscience and quantum field theory makes this seem to be not possible, and old-time religion no longer works as an explanatory of reality as we construct it, devise it, participate in it and even confuse it. The evidence is simply too great. To persist is to hold the potential for delusion, not illusion or fiction, and lead to but neurosis. This neurosis will be the inability to hold together ambiguities learnt from our now STEM society and our old religious minds/souls. This neurosis will lead to violence, self- and other-centered. Violence, particularly warfare and social/criminal, has been in decline numerically in the last half century in the West (it is disputed globally) and yet it seems that we are more aware that this is merely the manifestations of a social pressure gasket and the twentieth century may be set off as new culture wars that become civil or worse. And perhaps this is what we see in the last century of religion—a massive attempt to resolve the neurosis as we try to be "good" in the death of traditional theism. Perhaps fundamentalism—whether in religion or scientism—is the birth pains of a new mind? The final chapter examines whether there is a foundation to build on the new metaphysics of no metaphysics.

But, whatever, we hope we are decidedly in *terra nova*. Some (famously R Buckminister Fuller, R. Kurzwell and various tech groups such as IBM and Google) estimate that the advance of human knowledge about our physical world has doubled in the past 20 years and that this is only moving faster with the internet of things, artificial intelligence, and transhumanism on the horizon. This is the world birthed by science and its impact. One day soon, extraterrestrial life will be detected and perhaps even first contact. Our representation of the world—our illusions—will need to be continually altered at speeds previously unimagined. Are we ready? What role will traditional religion hold in this brave new world? One thing is for certain, this new world is not the world of the scientist simply told to shut up and calculate. But it is also not the world of the old soul either. It will, however, be our imagined world in the sense that we will need to make sense of it, order it and live in it.

Chapter 6

CAN WE BE GOOD? WHAT MODELS DO WE HAVE?

> Where questions of religion are concerned, people are guilty of every possible kind of insincerity and intellectual misdemeanor. Philosophers stretch the meaning of words until they retain scarcely anything of their original sense; by calling "God" some vague abstraction which *they* have created for themselves, they pose as deists, as believers [...] although their God is nothing but an insubstantial shadow and no longer the mighty personality of religious doctrine.[1]

Of the writing of many books in the last decades on the "fate of reason," "modernity" and theology's insipient contributions to their demise seem to have no end.[2] Equally the bell tolls for the university humanities and there is likewise a surplus of academic work on the transitory and contingent nature of what was once thought of as modernity's invariant principles of rational inquiry, now nakedly exposed as custom, hegemony and other pestilences and therefore openly hostile to any apposite "Other" in terms of economics, politics, race or gender.[3]

---

1 Freud, *Future of an Illusion*, 56.

2 Some of the better options include L. Dupré, *Passage to Modernity* (New Haven, CT: Yale University Press,1993); Michael Buckley, *At the Origins of Modern Atheism* (New Haven, CT: Yale University Press, 1987); F. C. Beiser, *The Fate of Reason* (Cambridge: Harvard University Press, 1987); Charles Taylor, *Sources of the Self* (Cambridge: Harvard University Press, 1989); C. J. Stout, *The Flight from Authority* (Notre Dame: University of Notre Dame Press,1981); E. Jüngel, *God as the Mystery of the World*, trans. J. Webster (Edinburgh: T&T Clark, 2001); Walter Lowe, *Theology and Difference: The Wound of Reason* (Bloomington: Indiana University Press, 1993); and Canada's own George P. Grant, "Research in the Humanities," in *Technology and Justice* (Toronto: Anaisi, 1986).

3 Again, the better options include M. Foucault, *The Order of Things: An Archaeology of the Human Sciences* (London: Tavistock, 1970); M. Foucault, *The Archaeology of Knowledge* (London: Tavistock, 1972); J. Guillory, *Cultural Capital: The Problem of Literary Canon Formation* (Chicago: University of Chicago Press, 1993); P. Bourdieu, *Homo Academicus* (Cambridge: Polity Press, 1993); P. A. Bové, *Intellectuals in Power: A Genealogy of Critical*

The landscape is well-mapped. This is the entire postmodern/postcolonial world in which Western academics, religion and life now live within. Interestingly, both religion and science are problematized but for different reasons.

The standard story is as follows. A nexus of complex interactions—in thought, geo and local politics, economics and religion—contributed to the rise of a way of thinking variously called the Enlightenment or modernity that assumed it possible to have an objective (and therefore impartial) point of critique that, if followed, promised, and has indeed given much, in terms of, human freedom, knowledge and even peaceable lives. This eristic promise ultimately placed the contingent individual self as singular, as a *common* anthropological reference to the nature of human and non–human beingness. From this new "self as ontological incarnation" of human ideality (and my choice of Louis Dumont's problematic term is intentional),[4] the human person, or a community of knowing selves (primarily as scientists but also as *Wissenschaft* in general), able to summon, judge and therefore discover the true nature of reality. Finally, we break the old world of the medieval and ancient to rebuild and find authority only in the products of our own hands. This, in turn, becomes the justification for all kinds of evils from racism to colonialism—all given the value of being justified in the name of scientific, universal progress. Science is the poster child here. It creates the industrial revolution, justifies expansion of territory for new resources and reduces human cultures to phenotypes. The humanities fare a little better, but only just so. It will soon be charged with the tacit claims of normativity for a ghost of Judeo-Christian concerns in gender and in sexuality.

Implicit above are a host of modernity's most precious icons, to which there has been regular reaction akin to Newton's third law of motion. These icons are (1) the ideal of absolute freedom-authority and its correlate that the self stands independent of culture-community and can rely on its own interiority to make true representations of reality; (2) the ideal of genius and its correlate that education is not merely a generic human enterprise but that *what* is taught is as important as that we are taught. What tethers the two icons

---

Humanism (Columbia: Columbia University Press, 1986); P. A. Bové, *Mastering Discourse: The Politics of Intellectual Culture* (Durham, NC: Duke University Press, 1992).

4 It seems to me that Dumont's contention that modern society places the individual, as the ontological incarnation of value, in the placeholder usually given to ideals as embodied in community, while certainly true, still falls prey to the Feuerbachian or Freudian analysis. In this instance, what can be said of the self as solipsist is equally prone in the "Other" of community. See Louis Dumont, *Homo Hierarchicus: The Caste System and Its Implication*, trans. M. Sainsbury (Chicago: University of Chicago Press, 1980).

together, the free self and the genius educator, is that the latter, the genius, is the best example of the former, the free self. The genius-avatar is the new priest or clerisy (to borrow from Coleridge) whose temple is the temple of the self and who baptizes everyone else into this temple through seemingly objective impartial education. These genius minds hold court in our public universities, indoctrinate others and the talking head of academia is the expert on all matters. The critiques of postmodernity rightly point out that there is no privileged view of reality and that what modernity was, in part, inhabited the realm of bias and privilege. Thus, as in the previous chapters, Freud's ideas fit quite nicely.

This is an admittedly bleak picture, and the university as general culture is split into the new tribes of social media, each tribe speaking to itself and its own kinship. The belief is that a common ground seems hopelessly lost, the hermeneutic of critique eating its own tail into fragments.

It is this ascent and zenith of the "transcending summonser" (George Grant) whose interior life dominates modernity and that promises too much, gives too little and is biased to its own bias. Instead, for many, what should be practiced is a postfoundationalism in both religion and in science, empirical and human.

The dominant flavor of this "transcending summonser" was and is instrumental rationality, and the golden child was and is the empirical sciences with their explanatory power and "magic" to transform our lives through technological advance. The modern university follows suit to this beloved child in all areas and becomes, regardless of discipline, the location of a common society of rational "scientists."

What did this mean? Let us parse this carefully. Rationality, and its critical *safe* dissemination (via procedure or method), assumed both that the world is objectifiable and that this tool stands outside context in order to impartially make representation. In short, the human subject as a knower or scientist becomes the location of meaning and correct method—namely, adhering to normative rules of praxis—promised a payoff. This payoff, so it was hoped, would be education in the virtues of tolerance and peaceable social life. The modern university, as a symbol of modernity, came to hold certain myths and the twentieth century has seen, in all disciplines including the once ironclad empirical sciences, those myths nakedly exposed. Most dominant, and still powerful, is the myth that the university is, by definition, iconoclastic, and therefore tolerant because said *Wissenschaft* is an *obligation* toward free inquiry and *toward* authority. Embedded, and problematic, in this notion is that particularity, whether ideological, cultural or religious, must be reduced or explained in terms of the general or normative to be "authoritative." Authority, and therefore forms of hard and soft policing, is recast as willful conformity to the process and outcomes of *Wissenschaft* itself.[5] However, the rise and proliferation of various movements or "isms" in all the humanities and social science disciplines have articulated this as exclusionary and artificial as a mode of authority as any it wanted to banish in the first place, including theology. In fact, and a person with the surname Marks must think this, authority so named was in fact another example of fetishism in which products of human hands or minds assume power over the originating community because the community, or powerful few or clerisy, tacitly hoodwinked us to believe it thus. Rather than inducing Nietzsche's creative multiplicity or Hume's tolerating spirit, the modern self is more arrogant and virulent than the tradition it exposed because it feigns neutrality while holding a new totalitarianism in all areas of human life with this reductionism.

Enter postmodernity and its "hermeneutic of suspicion," the persistence of critique eating its own tail, and an unwillingness by the "Other" to be subsumed into the homogenized *individuum* of modernity—the rejection of grand or meta narrative and all subsequent attempts to foist or construct another in its place.[6] Postmodernity in this rejection of supraorder and authority embraces paradox, fluidity and heterogeneity in favor of the ecstatic *difference*, secondariness and adopts a Nietzschian *Vornehmhiet* or nobility to will power or meaning—meaning is what we make.[7] Into this vacuum of postmodernity many rush where angels fear to tread.

At one level, science is immune to this critique—it is "shut up and calculate" as big pharmaceutical companies, the military–industrial complex and now communications fund it. It does not require public money and scrutiny—it remains captive to the market. Fundamentally, however, science is impacted as it is told that it has no metaphysic just utility and to leave such things to experts in terms of cultural norms. But, as this work argues, this is not possible. It is a human exercise, and therefore an education to reality.

Religion, and particularly the theological reflections of various religions, fell much earlier to the critique of modernity as it was one of the reasons modernity flourished in the first place. Religion carried in its bones the movement to the self as knower. As such, its history is much more useful to

5 See the following texts on this point: Kieran Flanagan, *The Enchantment of Sociology: A Study of Theology and Culture* (London: MacMillan, 1996); Steve Bruce (ed.), *Religion and Modernization: Sociologists and Historians Debate the Secularization Thesis* (Oxford: Clarendon Press, 1992).

6 J. F. Lyotard, *The Postmodern Condition: A Report on Knowledge* (Manchester: University of Manchester Press, 1986).

7 Who else but Nietzsche (not Heidegger as read through post-Heideggerians such as Levinas and Derrida) could represent a true theological (or rather atheological) postmodernity?

understand for science as eventually science will end in the same place—without a metaphysics and told to "shut up and calculate" while other ideas shape mores. There are at least three theological, although other disciplines have similar, positions or fashions adopted in response to this new postmodern world of dispersed authority and refused particularity. But the question is whether any of these positions works?

Theological Responses to Postmodernism

The most facile welcomes this erosion, as it levels the playing field for all while secretly thinking it has a leg up. I call it the "smiling sectarian." In this model, the call, at least for the university, is the reintroduction of multiple centers of dialogue so that, using Kymlicka's idea there is no one neutral space[8] and further, following Alasdair Macintyre's work (and likewise in the empirical and biological sciences)[9] that there is no neutral method of inquiry and therefore no real "higher ground." The correct mantra then becomes that true pluralism is served as in the process of discovering "us" we truly engage with "them" and therefore the floodgates of genuinely plural voices are opened. Thus, there a growing number of voices argue for the reintroduction of religious-based universities or general education. There is a measure of truth—the possibility of theologized identities—here, but also a danger that is far greater in my mind. The danger is that the only address to the "Other" in this case is a condemning or lamenting melancholy for the better days of moral standards and Judeo-Christian society or as a prelude to some apologetic or eristic work. Here, theologically, because the world is already ordered to God and what theology offers, usually in Thomistic tones, is a GUT of everything that cements, as in the past when it was the Queen of Sciences, all disciplines together usually around virtue theories of good and ethics. The argument goes that what needs to occur then is a rediscovery or rearticulation—a *reparatio*—of new commonplaces that have been in actuality all the time Christian (with a now postmodern tip of the hat to Judaism and Islam).

To my mind, this raises religion for the purposes of apologetic, encountering the same argument as when it was the Queen of the Sciences—in its haste to

---

8 See, for example, W. Kymlicka, *Politics in the Vernacular: Nationalism, Multiculturalism and Citizenship* (Oxford: Oxford University Press, 2001); Charles Taylor et al., *Multiculturalism: Examining the Politics of Recognition* (Princeton: Princeton University Press, 1994); Charles Taylor, "Democracy, Inclusive and Exclusive," in *Meaning and Modernity*, ed. R. Madsen et al. (Los Angeles: University of California Press, 2002), pp. 181–94.

9 Alasdair MacIntyre, *After Virtue* (London: Duckworth, 1985); MacIntyre, *Whose Justice? Which Rationality?* (London: Duckworth, 1988); MacIntyre, *Three Rival Version of Moral Inquiry* (London: Duckworth, 1990).

speak, correlate and engage—does it fail to take its cue from a more central purpose, which is theology, for the sake of worshipping community? It is not without some irony that I note the most able commentator of this position—Gavin D'Costa—argues that the villain of apologetical *ratio*, whether in Aquinas, nominalism's Ockham or the midpoint in Duns Scotus—*did* in just such a model become disengaged from its home in Christian commitment once before.[10]

If we see a precursor here for science, it must surely be the old sawhorse of the positivist materialists who argued that science has no metaphysics as it cannot entertain those questions de facto. It is told to "shut up and calculate" while its utility and meaning is left to other educators to reality. This is the sadder history of science, the boogey man of literature with its Doc Brown scientist who fails to understand the consequences of his actions.

The next model, I call "kicking it old school." Model two is slightly better and drives home the above tacit thesis that Christianity (or another world religion) *because* it has a metanarrative *that has* worked can now offer itself as the prodigal son of modernity whose value as its *moral* memory and *house* religion is now only recognized as such. In this case, theology offers a response to the seemingly evacuated moral center of modern life or it gives a common discourse that represents the "awe" of intellectual and human life. Theology or religion in this postmodern academy functions as a theoretical conscience—a Jimmy Cricket for modernity's wooden Pinocchio—and gives a common spiritual-linguistic heritage from which to play—that is, it plays a *sociological* function of giving identity and value to Western culture if respected. Theology morphs to functionality and religion is back on the table, if only because there is "bonding capital" to its reintroduction in public life.[11] This, at least in my mind, is superior to the first model because it formally and explicitly renders theology functional—that is, to reinforce it as a sociological construct of value. This sits well with not only the academy but also sociopolitical understandings that religion is useful if it remains private, and when public it must function as a social glue or human capital. It is honest in its unveiled appeal to its authority but conflates that authority with itself. This position, I note, is also similar with Freud's comment that religion can give some social structural and

---

10 D'Costa gives a fair summary of the literature. Gavin D'Costa, *Theology in the Public Square: Church, Academy and Nation* (Oxford: Blackwell, 2009), pp. 9–13.

11 The objectification of modernity's "gaze" that things become evaluated as to their end. See Michel Foucault, *The Order of Things: Archaeology of the Human Sciences* (London: Tavistock, 1974). This is also the position of some other nontheologians such as John Biles and I would contend the *raison d'etre* of the politics of representation fostered by Charles Taylor.

therefore personal stability, and has, in fact, done this well and suitably in a prescientific age of understanding.

This is very close to Freud's proposal, or that of the Vienna Circle in terms of how to work in macrosystems of economics, politics and so forth, by adopting a purely empirical and positivistic approach. Following the work of Mach, Wittgenstein and other associated, these scientists argue for evidence and couch it carefully as linguistic expressions of fundamental reality, which is an asymptote. It is to be replaced by new discoveries and evidence and it is an education toward reality. It rescues us from ghosts, goblins and Popes whose decisions are based on old false science. The issue here is that its insistence on verification and language destroys its foundation of science. It is impossible, in other words, to not make metaphysical claims that exceed the limit of what can be said. It is too immersed in modernity's confidence on human knowing as being privileged. Like its theological cousin, it mainly works because it helps to address the most egregious of problems of human knowledge based on superstition and ignorance.

Finally, the third position—the "reconstructionist"—is the most nuanced and bears the most attention being paid to because it is seductive to religious minds. It also, I think, bears the most similarity to Freud's critique of the inherent insincerity and intellectual dishonesty that religion as illusion as quoted at the beginning of this chapter. The third position is the thesis that postmodernity has "re-enchanted" or "remythologized" the world with a new sacramental cosmology and that in this new space radical Judeo-Christian identity and authority can flourish once again.[12] The apparent *apotheosis*, or divinity of the human self, while remaining true to modernity's hypercritique, becomes the project of this school. In short, how can one deconstruct and still hope for a world of mystery? The answer is a spirituality of secondariness—Derrida's *khora*.[13] The *khora*, in refusing definition, becomes the leitmotif of re-enchantment as one need to come sideways, backwards and randomly at spiritual things in order to avoid system and abuse; and find in the ecstatic or unusual, a moment of fleeting purity. The *khora* shows all formal things are unstable, externally forced and embedded in human activity and thus serves through its very opaqueness to remind us that the ecstatic "more" is "realer" than what we know: "There are more things in heaven and earth, [modernity], than are dreamt of in your philosophy."[14] The world, accordingly, is suddenly

12 For a fuller treatment, see Graham Ward, "Postmodernity and Protestantism," in *Blackwell Companion to Protestantism*, ed. A. E. McGrath and D. C. Marks (Oxford: Blackwell, 2003), pp. 453–67.

13 Jacques Derrida, *On the Name*, trans. D. Wood et al. (Stanford: Stanford University Press, 1993).

14 W. Shakespeare, *Hamlet*, I, V, 174–75.

re-divinized, all is mystery as an account of ontotheology, abuse by religion of spirit, is forbidden by definition. As Levinas notes, the sacred is the incessant white noise of the universe bleeding into every determinate sound all the while defying attachment to any sound except as its deconstructed absence.[15] Suddenly, everything is spiritual and everything open to being repaired if only we are willing to look in new places—but isn't there an old saying about new wine and old wineskins? But this is the stuff of art and song, and it is very powerful as who wants to quantify love?

The most serious issue for the reconstructionist, and why this does not get much traction out of popular books and educated theology students, is that by resisting any framing or taxonomy except an ever-radical structure of critique to authority and particularity this sacred must, by definition, be always and forever anti-institutional and hostile to community—it therefore must occlude the very identity that it hopes to provide. Side-stepping ontological and revelation issues for now, this means that dominant Christian (or Jewish) forms that serve to orient Christians (or Jewish) to the world—baptism, Eucharist, common Church life, catechism, bible reading, sermon hearing, public worship and actions of charity—must all be rejected as being oppressively sociologizing. These cannot be locations of formation or a place—even given divine sanction—from which we are sent back into the world. Further, and more difficult to follow, the postmodern theological shift of agency within the "sign" from God's *agape* to human *eros* stresses not divine action "in, under and through" praxis but rather refers to the potential of *our* desires or spirit as being elicited through opaque, random signs. Religious common life is one discovering spiritual resonance in and to the symbol of the community but with the priority of the self. In this case, the recovery of *eros* over *agape* carries its own destruction, while we make meaning, this meaning must always tend toward nihilism because it is, by definition, as nonteleological as its source. The only primary is secondariness[16] and this must lead to a sophism not only in terms of logic but more importantly in building a common life. Freud, again, is helpful with his pithy remark: "Critics persist in calling 'deeply religious' a person who confesses to a sense of man's insignificance and impotence in face of the universe, but it is not this feeling that constitutes the essence of religious emotion, but more important, the next step, the reaction to it which seeks a remedy against this

---

15 See E. Levinas, *Time and the Other*, trans. R. Cohen (Pittsburgh: Duquesnue University Press, 1987).

16 The noted Hindu scholar David Smith has argued just such an alien imposition, as well as tracing efforts to place Hinduism into the second camp given, on Hindu studies; David Smith, *Hinduism and Modernity* (Oxford: Blackwell, 2003).

feeling?"[17] For Freud, the entire system of theology as community was to make this feeling go away, to give civilization and culture its role in socializing us. To do away with this, and remain "religious" is "irreligious in the truest sense of the word."[18] In short, the insistence on religion clears out any use for religion except to atomic experiences of one self. That is surely lonely?

The scientific relative to this position is the emergence crowd, the pantheistic Bohm endorsing, multiverse-enamored scientific metaphysicians who see in the claims of science and religion the same undergirding of divinity. These are the scientists and theologians who claim that postfoundationalism must, by necessity, be the ground for a new age (pun intended) of science and theology/religion in dialogue. They claim the failures, like their theological relatives, of both prior responses point only to one endpoint—new reconstruction of human knowledge. Deepak Chopra and the failure of the Copenhagen Interpretation as totally expressive of all reality both cheekily point to the same thing—quantum healing.[19]

Theology, so the tale as told, is the first victim in the move to interior uniformity and displacement in the modern research university. The reason for this is that theology, classically, always took seriously the notion of "Otherness" in a manner that refuses modernity. The proposition that God "is," for theology, and therefore that human creatures are under authority, was a consideration in its discipline. This could have functioned to decenter the wider project of modernity and its center in the human person but did not. Theology's historical response, however, was disastrous and thus left theology open to the devastating critique of Marx, Feuerbach or Freud or any of their progeny. Theology it seems then camps on the moral—"as long as [people] do not discover that people no longer believe in God, all is well […] [but] if you then discover that there is no God and one need not fear his punishment, then you will certainly kill without hesitation […] and so follows the necessity for either the most rigorous suppression […] or a fundamental revision [of the idea of God].[20]

A Postfoundationism or a Reconstruction?

In the past decade, the above problems for religion (and by extension, the wound of reason as *wissenshaftliche* and therefore also science), has opened a double critique of historic science and religion as both being ultimately

17 Freud, *Future of an Illusion*, 57.

18 Freud, *Future of an Illusion*, 57.

19 Becker cites a *Futurama* episode that makes this point. Becker, *What Is Real?*, 281.

20 Freud, *Future of an Illusion*, 69.

foundational. The primary thinker on this front is J. Wentzel Van Huyssteen. Van Huyssteen's premise is simple. Just as religion needed to move away from foundational exclusive statements on the Real, and to learn that divinity may be more complex than one singular *tradition* (and use different academic tools that dot the landscape of theological training now from critical theory to archeology to psychology), so has science from its classical form in which it is the only game in town to describe all of reality. The poster child, however, is not quantum mechanics but evolutionary biology, and the process of self-consciousness.[21] Van Huyssteen's proposal is that science, given the revolutions of post-positivism, post-Kuhnian and post-historicist critique, is now baldly aware that its narrative is human driven, human bias and human infected[22] and that this is to be expected as we are part of the process that created us in the first place.[23] In short, we are observers in our own system, and this means humanity was reinserted into nature so that epistemology is all we are. There is no duality in us, we are, simply, in this world, including the world of our minds and souls.

The power of the proposal is that it promotes humility, interdisciplinary hearing (including across the divide) and works primarily as a place holder. Van Huyssteen notes that it may be hostile to traditional theism, and on this, he is correct as it is pantheism or process theology. It is the reconstructionist position based on epistemology and exploiting the fog of consciousness and the persistence of religious belief:

> I will argue, however, that although this may explain away deistic notions of God, it does not fully explain religious belief and certainly does not explain away belief in God as such.[24]

Van Huyssteen does engage with Freud (albeit through an intermediary Franz Wuketits) and the claim that religious consciousness, including souls, sprung from the historical positive situations of primitive cultures in a prescientific age as coping with specific life conditions.[25] Van Huyssteen turns Freud seemingly on his head. It is not the origin of belief, but that very persistence that needs to be accounted for in any conversation between science and religion. Our evolving brains (and,

---

21 J. Wentzel Van Huyssteen, *Duet or Duel: Theology and Science in a Postmodern World* (London: SCM Press, 1998), pp. 130–42.

22 Van Huyssteen, *Duet or Duel*, 15.

23 Van Huyssteen, *Duet or Duel*, 135.

24 Van Huyssteen, *Duet or Duel*, 135. Note he is not referring to Deism in its classical form per se, but some analog that assumes a divinely aware personal being that is connected to the world rather than the disinterested first cause of Deism.

25 Van Huyssteen, *Duet or Duel*, 154.

again, presumably of other hominids) are selected to make religious belief and that Wuketits (and the other New Atheists such as Richard Dawkins) confused the environment with the process. Evolution selected brains that could make religious beliefs, not harsh environments created brains that confused natural things for supernatural realities? Van Huyssteen seems to both disagree and agree with Freud in one sentence. On the one hand, he seems to argue that cultural evolution is nature plus and that ideas such as religious belief that manifest concretely exceed naturalism. This is very similar to the concept of emergence with consciousness. But he also argues that evolution—the hardwiring or wetware physical brain—is exactly natural. But the implication is that at some point, some magical "pixie-dust in the microtubules" manifests as the natural becomes the creative or spiritual brain.[26] Where he agrees with Freud is that cultural evolution is human made and not ontology but epistemology, but then he gives it an ontology (creeping dualism back in) as an emergent property that is religious impulse:

> Thanks to a profoundly post-foundationalist argument of evolutionary epistemology, we are now able to move beyond the point where our only option is a forced choice between resolute naturalism or radical supernaturalism. Belief in God should be an *immediate* and *natural* interpretation of our *experience*, communicating an *underlying personal* reality which is *like us* in some very fundamental aspects.[27]

But note the italicized words. Everything is subjective, it is clearly personal and extrapolation of mind/epistemology back on to reality. It is fiction, it is illusion and perhaps it is delusion. In this case, it is delusion because it leads exactly to Keith Ward's panentheism,[28] which Van Huyssteen quotes thus:

> If the goal of the cosmos is to generate consciousness, and if its creator is the supremely conscious agency of God, then the realization of this goal must include a knowledge of the existence of God, of the dependence on the cosmos on God, and of the likeness of mental existence to divine existence. In other words, the goal is likely to lie in the realization of a conscious relationship of the cosmos to its creator.[29]

---

26 Van Huyssteen, *Duet or Duel*, 157.

27 Van Huyssteen, *Duet or Duel*, 160, emphasis mine.

28 I spent some time with Keith Ward at Oxford. His theological pursuits are captured in my own work: "This is that there is a spiritual realm which is in some sense more real than, and probably the basis of, the material world." Darren C. Marks, *Shaping a Theological Mind* (Aldershot: Ashgate, 2002), p. 127.

29 Van Huyssteen, *Duet or Duel*, 160.

While there is some fruit in Van Huyssteen's proposal in that it makes religion pay attention to science, it fails to address the most important distinction between the two disciplines, and one raised by Freud himself. Only science can take at its trust value that it is entirely natural, and it checks itself constantly to that claim. Only science realizes that epistemology is not ontology, and that ontology is found through a focusing of skeptical questions as we educate toward reality.

Van Huyssteen's postfoundationalism is crypto-apologetics for religious belief, and in particular, the belief in a moral universe. But if it leads to religious nihilism, what does that mean for God and the soul as a unique human expression of that divine relationship? It is, as in the work of Haidt, a question of morality or common life, always the reduction of the modern religious experience to the moral concrete world. The question becomes, without God, can we be moral.

A Morality of Sentiment

The question of whether we can be moral, or humane, without God is the worry that religion-loving, nature-endorsing minds, postfoundationalists and a host of others have. And the specter of scientism, and of science, haunts them as science undermines the myths or illusions of control, uniqueness, progress and responsibility that traditional religion, and Judeo-Christian traditions, have given.[30] Into this vacuum, in part created by science and its insistence that we are merely self-conscious animals, the past centuries have seen numerous new myths—capitalism, liberalism, Marxism, socialism and consumerism to name but a few—but all these new myths share a belief that progress and goodness are attached to a future goal, a hope and a redemption from our worse animal selves. Even atheism, according to John Grey, is a stop gap position—it merely set the human person rather than God in the position of transcendent control and in the self-belief of human activity. Divine will, order and purpose become human will, order and purpose and the attributes thought of as divine become part of humanity and its Realpolitik. Instead of God redeeming, we redeem ourselves.[31] To this challenge, theism and traditional religion correctly point out the privation of the position, and humanity cannot remain naturally natural. We are, of course, inveterate fiction makers. And the history of the past century seems to lead to the conclusion that we are poorer, morally and in terms of generosity of spirit as a result. And the leading teacher in this

30 John Grey, *The Silence of Animals: On Progress and Other Modern Myths* (New York: Farrar, Strauss and Giroux, 2013), p. 82.

31 Grey, *Silence of Animals*, 131–45.

process is science, which replaced God with atheism (or positivistic scientism) or postfoundational panentheism in our postmodern condition. It seems that without God, history tells us we are morally bankrupt, and that at best, the universe needed humans to think about themselves to make sense of it and to give it purpose, shape and morality.[32] If God did not exist, we would need to dream up God to quiet our restless minds, make order and give purpose.

But is this necessary? It may be true, but is it necessary? There are two things to consider in answering this question. The first is whether it is possible to construct a morality out of something that does not necessarily require notions of transcendence as the foundation for being or making moral decisions. The second is whether being an animal rather than a demigod is helpful for humanity in its future.

To answer the first consideration, we can learn from Western history's first great modern skeptic/atheist David Hume and his morality of sentiment. Hume's entire project centered on the rejection of any claim that human persons have divine-like capacities (with various sources from souls to panpsychism to minds) on the ground that any analogy between a seemingly infinite omnipotent being and humans is unwarranted. Belief in this is just exactly that, pious belief and cannot be trusted (and certainly experienced as such) as a guide to human betterment. In Freudian terms, it is a fiction that causes delusional actions. To answer this, Hume suggests a morality based on sentiment that we may better call empathy in our context. This empathy or sentiment is not purposed (i.e., not part of some divine plan that needs redemption, correction or intervention), is not transcendent as a special uneffaced part of human constitution such as a soul or even rational mind and is tied to what is the most common human experience in emotion itself. Hume, then, suggests finding a morality based on the human experience, read as emotional response, of conflict—external and internal. What is "transcendent" is that humans *feel* something constantly between the external world as what it is, and what they would like to feel the world to be. His morality is full of phantasms, shadows and ghosts that we all feel acutely, and even erroneously. This feeling is our desire for meaning, and it is local and singular. It is the desire for meaning that is universal, and this is what Hume gravitates toward.

For Hume, empathy is generated by analogical association that holds two modes in tension. The first is the presentation of an objective other before the self and which the self "learns" associations including moral associations of pleasure and pain as external to the self (e.g., this situation caused pain or pleasure in that person). The second is the self-presentation within by which

32 Grey, *Silence of Animals*, 164.

the self learns to associate moral categories internal to the self as consequence of the former experience (e.g., this situation of pleasure or pain in that person caused me to reflect on my own response of pain or pleasure). This, to the alert reader, should sound very Freudian with the ego/id being the self and the superego being the experience of the moral Other before me. For Hume, the process of reflection—learning to balance self-interest with self-awareness and other awareness—is the work of empathy. It is personal and, by necessity, communal and it is also simultaneously messy and exacting. The presence of the morally demanding other before me causes a response, not a purely rational but an emotional–rational conflict and that conflict leads the agent to moral meaning. This moral meaning, in turn, is tested in the community where it is received, judged or rejected, and this, in turn, causes further testing emotionally and cognitively. But it is always, first and foremost, exactly a conflict of meaning making, desire and rationalization. There is no absolute source of authority, no end to the conversation because it is God's preordained *Logos* hidden in some special capacity of human speciesism. It is our self-consciousness itself. It is humans coming to terms with their limited roles as moral agents, and this burden makes us empathetic—open to correction from the moral other as well as terrible in our exclusion when we confuse categories by thinking we are perfected or given privileged judgments. That, for someone like Hume, is delusional.

Hume, and those who follow this morality of sentiment, is not naïve and thinks this is done cheaply or easily. It is costly and difficult, requires humility and a careful accounting and system of government that promotes such chaos as formation of moral sentiment. But most of all, it requires the loss of the most precious gift of religion—the self-belief of human uniqueness and its concomitant sources of that uniqueness in souls, minds or wills. It requires the death of redemption, the category of forgiveness as determined by the Judeo-Christian mind and the loss of purpose.

Instead, sentiment tells us that we do not need redemption but courage to end our misery. In this Marx was correct; religion harms us because it tells us that redemption is more important than resistance and change. John Grey wrote that in the Christian mind the tragedy is not that Jesus died, but that he rose. There is, for him, no redemption from being human as all we are is human.[33] Being human is to make fictions, cognitive-emotional fictions within and without, and then to act on them as they present moral decisions on us. Sometimes, perhaps far too often, this will mean terrible societies and sometimes not. But it is "education to reality" rather than delusion dressed as that cognitive-emotional fiction in divine clothing and power.

---

33 Grey, *Silence of Animals*, 203.

A morality of sentiment follows Freud's work quite nicely. The goal of moral reasoning is the cessation of conflict caused by the confusion within with the confusion of the world without. It is either making peace with it by acting accordingly or living in anxiety and delusion. Like psychotherapy, the goal is not perfection but self-consciousness itself.

Poetic Naturalism or Nature Realism

So, what is the power of science then, and of neuroscience and quantum physics, in this project? It is what Sean Carroll calls poetic naturalism[34] and I have called nature realist—the decentering of humanity as special due to divine fiat and the placement of humanity in nature as natural with its unique self-consciousness of being in conflict morally. We cannot help but make fictions—we were accidentally selected to be this by evolution because we are brain all the way down—but we have at present enough understanding of the core theory that we are aware that our old fictions—religion itself—cannot tell us truthfully about nature, our nature and therefore our fictions. We need new fictions, new commitments that guide us in our conflicted self-consciousness (culture and civilizations) that no longer come from our Stone Age souls but rather our contemporary minds. Science tells us we are merely natural—that our self-consciousness is merely a manifestation of biochemical processes and this is prone to error, damage and manipulation because of the quality of the machine itself. Science tells us that our uniqueness and divine purpose therefore is nothing more than a local expression of physical parameters, that we are not truly unique, and that the universe is uncaring, cold and hostile to us. Science tells us that this is the only existence we will have, that our bodies—which is all we are—cannot exist in some future resurrection. Finally, science tells us that there may be many universes, but all universes will be local, and that the heaven of God (or God of heaven) is merely a construction of delusional minds. All there "is" is a universe. We are blips in a universe that has always and will always be larger than we are or can understand and will one day even end. We are local to our planet, unique only to it and therefore closer in kinship to our coinhabitants than angels or other celestial beings. Christ Church's world of demons, angels and humans is replaced by the University of Oxford's Museum of Natural History's statuary of primates, mammals, and single-cell organisms.

What this all means, at least to this person, is that when we are stripped away from our illusions or delusions of religion and exposed as shallow meaning

---

34 Carroll, *Big Picture*, 289–90.

making inveterate fiction makers that we stand at a crossroad. This crossroad is to fit our moral sentiments along those truths—that we are one with this planet and not a divine home and that we share more with others here regardless of economics or politics than divine beings—and to construct a morality that acknowledges the fickleness of our illusion making moral sentiments because they are imperfect self-representations of our brains collapsing core field theory constants. This is messy, dirty and heuristic. It does not stand privileged. It is personal and communal. It will be abused (but steadfast defense of the basic illusory nature guards this). But it is also the best hope for continued existence on this planet, for making decisions that acknowledge the fragility of our ecosystem, protects the burgeoning population without massive culls and seeks to expand human life outside this small planet while protecting the naturalness of other people's existence as lifestyles and gender and so forth. It may be true that we cannot remain naturally natural, but science provides us "education to reality" that helps to offset our most defensive anxieties that we transform into God and God's will. Freud is correct in that science is our best hope, not because it stays out of metaphysics, but rather because it corrects fictions therein.

Toward a Theological Theology

Paraphrasing Kant, we may live in an age of science, but not yet in scientific age as we still have massive influence from religion. Therefore, a legitimate question is raised by Van Huyssteen—how does theology or religion relate to science? Rather than his proposal that both disciplines seek a common field, I would argue for a theological theology to be practiced. The value of this is that it recognizes the local nature, and therefore the extent of the field to which it must yield to a dominant scientific world, of theology.

What do I mean by a "theological" theology? By theological theology, what I am arguing toward is a reformation (again the word choice is purposed). This involves several considerations, all of which serve the faith community not by becoming utilitarian—theology for the purpose of function per se—but by helping *this* community in *this* time articulate its creative fidelity to its tradition.

The first step, and in most general terms, theological theology must take is a healing of its own fragmentation within its theological cognates, each of which has in various degrees and to various extents become enamored with contingent methods and alien forms. This "Great Reversal" (to coin an oft-used phrase of Frei) is not peculiar to our time, but what is peculiar to us is the new depth of split between exegesis of the Scripture and so-called practical theology as the performance of the Church/Synagogue in response. In simple terms, what has the ivory towers of theological education have to do with the lives of those who participate in religious life in the houses of worship? Theology students are largely left to place the random scrabble bag of consonants and vowels into a theological vocabulary on their own and leave "religiously illiterate" in terms of theology and rich in critique and accommodation. What I mean here is they continue to propagate the "house religion" of modernity, as models two and three above, and then, of course, scratch their heads when either their parishioners flounder or other Christian/Jewish theologies or communities (and the Global South) break step or communion because we appear strangers to them and they to us.

Theology must understand itself as not a free science but one controlled by its own particularity, tethered to its text in the Hebrew and Christian bibles (Scripture) and articulated in its account of Jewish-Christian history (tradition). All this is a local *account* of God, truly theology, in which is posited a hermeneutic, even a circle, that avoids the alien classifications of sociohistorical or other theory in order to link Jewish-Christian Scripture[35] and Judeo-Christian praxis to their function of witness to the presence of God in the created universe, redeeming and justifying human and created existence through this human institution called Church/Synagogue.

In other words, theology must remain theology. It cannot leap to the other world of science, to find as Van Huyssteen argues an implicit coconspirator. By remaining theological theology, it feeds its own communities, and perhaps, as the world turns, change happens because those fictions are found inadequate. Perhaps not. But theology is not science, and science is not theology. In this sense, religions should remain fundamentalist—this is a given choice and can be rejected as such or not on the grounds of the utility of the ideas carried within. I think it avoids the problem given in the three models that promote various forms of religious illiteracy, which propagates religious illiteracy in its own field and in doing so raises illegitimate and harmful relationship of authority. A theological faculty of theology in my model works from its literacy, creating a kind of theological penumbra—in which it is challenged and invites others to true dialogue, not the crypto-apologetics of Van Huyssteen and Ward whose vision is that of the populist meme of Jastrow's scientist arriving at a GUT only to see God sitting there, reading from the religious texts of our Stone Age ancestors.

---

35 Again, a litany of literature on this point can be sighted: N. Lash, *The Beginning and End of Religion* (Cambridge: Cambridge University Press, 1996); H. Frei, *The Eclipse of the Biblical Narrative* (New Haven, CT: Yale University Press, 1974); S. Fowl, *The Theological Interpretation of Scripture* (Oxford: Blackwell, 1997).

Theology, speaking as theology, then reminds science that what is truly *important* is not what is *neutral* or easily reduced; that the logic of *Wissenschaft* is not at all unbiased and therefore more important but rather that what makes learning for the purpose of common life useful is academic nonconformity and the rejection of commensuration. Theology, by appealing to its "orthodoxy" rejects the standardizing orthodoxy of modernity and answers its own apparent forms and in doing so reminds science of its intention through critical dialogue and questioning from differing readings of reality to promote a common life. Theology is not a vaccine for the ailments, that is too proprietary and arrogant, but a possible nutrient. Theology, however, must also realize that science in its torturing reality has this already built in, and that its growing pains came because its source in theology is problematic due to its origins. Theology is an education in an illusion *of* reality, it is not the same as a fiction as education *to* reality because it cannot leave its source of human-centeredness. It is not naturally realistic.

Many involved in science will part with me here. I recognize a present value for theological reflection on religion because as a working fiction it helps to ask the best kind of questions. We are not ever to approach the Vienna Circle manifesto promising to rid human inquiry of metaphysics and theologizing thought because we, including scientists, are inveterate idolaters in that we cannot help but make our observation of the natural something that means something else to us. We are to be natural realists, not endorsers or skeptics, but to guard against this intrusion as we educate ourselves to reality. Theology, historically, reminds us to include others in our sentiment of empathy as we find ourselves in a wider web of life in the universe. It can remind us that we can have too much pride in our uniqueness, that our fall to this belief is a metaphysical isolation and exile and that in the past, religion helped us to navigate this isolation. And, of course, this also means that there is not true dialogue but conflict between the two fields. One was useful because it revealed a natural realism, or decentering of human uniqueness—by exploring the very idea that humans were unique but not alone in that they answered to God and the divine powers. They do not share an ontological truth, except the truth that humans make up stories about themselves and we need better means to understand the impact of those stories.

Chapter 7

CONCLUDING POSTSCRIPT OR WE'RE OKAY?

> We're always, by the way, in fundamental physics, always trying to investigate those things in which we don't understand the conclusions. After we've checked them enough, we're okay.
>
> —Richard Feynmann

But if we are not quite ready for an age of science, and still are haunted by the ghosts of religion (and the Judeo-Christian world) then there are two places, again suggested by Freud, to work. The first is to recognize that "theological theology" is first and foremost necessary if there is to be any engagement with the scientific world. This has two reasons: (1) It allows for a critical evaluation of the role of socialization within its own traditions and can see whether this is either a resource or a hindrance rather than the reconstruction found in religious studies as a crypto-theological science. (2) It speaks its own language at the table of society as a result, and this, in part, also means that it must learn to NOT speak when its language is no longer fruitful for the conversation. Two weak examples suffice here. Clearly, the Judeo-Christian resource on sexuality in distinction of nature is different than what we scientifically know on sexuality and choice. Punishment or cultural restrictions on sexuality as nature then seem to be absurd as if we are only nature, then we cannot easily limit this behavior as expressed. We must account for (and this is indeed a brave new and terrifying world) behavior as a manifestation of nature at some fundamental level in both our social mores and social laws. Second, consider climate change denial and its relationship—undeniably so—to the Christian worldview of the dispensational mind that treats this present earth as a temporary place while waiting on the rapture. Whatever planet remains, will clearly for the dispensational mind (and this is a powerful voting body in the United States and elsewhere) be left for the sinners needing either chastisement (the Jews) or punishment (everyone else) before the end of the cosmos as we know it.

In both the above cases, the reality of social politics facing the global world is decidedly pointed in another direction, and yet, the religious viewpoint remains salient and powerful. Likewise, considerations on marriage, death, parenting choices and so forth—all under the freedom of religion—may as we learn about ourselves as natural, be in increasing conflict with religious ideas as understood. Add to this our present scientific understanding of addiction, impulse control and even sexuality and so much of our society will be altered from the criminal justice system to understanding how we create inclusive social networks. This is "our reality," and the question of how we understand our education to this reality is paramount. And of course, this excludes science fiction concepts still to come—extraterrestrial life, artificial intelligence, and transhuman augmentation. The grounds of theological self-understanding are already badly challenged by those ironclad scientific understanding of the brain. In these above instances, it may not be what religion can offer at all, but simply that religion cannot offer anything at all except an archeological tool in the same way we moderns view ancient Egypt and its pantheon. Religion, so interwoven in our minds, cannot speak to our new world. It does still, but it is rapidly shifting.

And we turn full circle back to our starting point of neuroscience. Neuroscience tells us we are "brain all the way down" and that whatever human self-definition we may have to come in the next century, will undoubtedly come from this knowledge and that all attempts to find divinity within our consciousness, our nanotubules of CQFT is really just a legerdemain—a magician's or philosopher's misdirection to keep the nature-endorsing metaphysics at odd with the nature skeptics—and our civilization will need to do the hard work of reconciling these ideas with our cultures. To answer the question of whether religion can answer neuroscience? The answer is "no." That is terrifying and we now need to work on asking whether we can find a peaceableness in this "silence of animals" as John Grey suggests. We are indeed fiction-makers, we construct illusions to answer our deepest needs and questions, and often this is exploited. But we are also in transition from delusion of religion to illusion self-understanding. Theology and religion still have a role, to parse itself as best it can as a true conversational partner and not a shadow self of deeper delusion and this is valuable and necessary. But it must also open itself to evidentiary claims, to asking whether the illusion is delusion, or whether a better illusion—often provided by science with all its warts and pimples—can work toward peaceableness. Again, Freud has the last word:

> I grant you the uncertainty of all these things. But we cannot refrain from saying what we think. [...] From this bondage [to delusion created

> out of fear of chaos] I am, we are, free. Since we are prepared to renounce a good part of our infantile wishes, we can bear it if some of our expectations prove to be illusions.[1]

In other words, we're okay.

---

1 Freud, *Future of an Illusion*, 93, 95.

Lightning Source UK Ltd.
Milton Keynes UK
UKHW012248191021
392489UK00001B/65

9 781839 980091